Beginning Apache Spark Using Azure Databricks

Unleashing Large Cluster Analytics in the Cloud

Robert Ilijason

APress®

Beginning Apache Spark Using Azure Databricks: Unleashing Large Cluster Analytics in the Cloud

Robert Ilijason
Viken, Sweden

ISBN-13 (pbk): 978-1-4842-5780-7 ISBN-13 (electronic): 978-1-4842-5781-4
https://doi.org/10.1007/978-1-4842-5781-4

Copyright © 2020 by Robert Ilijason

This work is subject to copyright. All rights are reserved by the Publisher, whether the whole or part of the material is concerned, specifically the rights of translation, reprinting, reuse of illustrations, recitation, broadcasting, reproduction on microfilms or in any other physical way, and transmission or information storage and retrieval, electronic adaptation, computer software, or by similar or dissimilar methodology now known or hereafter developed.

Trademarked names, logos, and images may appear in this book. Rather than use a trademark symbol with every occurrence of a trademarked name, logo, or image we use the names, logos, and images only in an editorial fashion and to the benefit of the trademark owner, with no intention of infringement of the trademark.

The use in this publication of trade names, trademarks, service marks, and similar terms, even if they are not identified as such, is not to be taken as an expression of opinion as to whether or not they are subject to proprietary rights.

While the advice and information in this book are believed to be true and accurate at the date of publication, neither the authors nor the editors nor the publisher can accept any legal responsibility for any errors or omissions that may be made. The publisher makes no warranty, express or implied, with respect to the material contained herein.

Managing Director, Apress Media LLC: Welmoed Spahr
Acquisitions Editor: Joan Murray
Development Editor: Laura Berendson
Coordinating Editor: Jill Balzano

Cover image designed by Freepik (www.freepik.com)

Distributed to the book trade worldwide by Springer Science+Business Media New York, 233 Spring Street, 6th Floor, New York, NY 10013. Phone 1-800-SPRINGER, fax (201) 348-4505, e-mail orders-ny@springer-sbm.com, or visit www.springeronline.com. Apress Media, LLC is a California LLC and the sole member (owner) is Springer Science + Business Media Finance Inc (SSBM Finance Inc). SSBM Finance Inc is a **Delaware** corporation.

For information on translations, please e-mail rights@apress.com, or visit http://www.apress.com/rights-permissions.

Apress titles may be purchased in bulk for academic, corporate, or promotional use. eBook versions and licenses are also available for most titles. For more information, reference our Print and eBook Bulk Sales web page at http://www.apress.com/bulk-sales.

Any source code or other supplementary material referenced by the author in this book is available to readers on GitHub via the book's product page, located at www.apress.com/9781484257807. For more detailed information, please visit http://www.apress.com/source-code.

Printed on acid-free paper

Select support from family;
SUPPORT

Malin

Max

Mom

Table of Contents

About the Author .... **xiii**

About the Technical Reviewer .... **xv**

Introduction .... **xvii**

Chapter 1: Introduction to Large-Scale Data Analytics **1**

   Analytics, the hype... 1

   Analytics, the reality.. 2

   Large-scale analytics for fun and profit.. 3

   Data: Fueling analytics... 5

   Free as in speech. And beer!.. 7

   Into the clouds... 8

   Databricks: Analytics for the lazy ones.. 9

   How to analyze data... 10

   Large-scale examples from the real world.. 12

      Telematics at Volvo Trucks.. 12

      Fraud detection at Visa... 12

      Customer analytics at Target.. 13

      Targeted ads at Cambridge Analytica... 13

   Summary.. 14

Chapter 2: Spark and Databricks **15**

   Apache Spark, the short overview... 15

   Databricks: Managed Spark... 17

      The far side of the Databricks moon.. 18

   Spark architecture.. 19

      Apache Spark processing... 20

TABLE OF CONTENTS

Cool components on top of Core 24

Summary 24

Chapter 3: Getting Started with Databricks 27

Cloud-only 27

Community edition: No money? No problem 28

    Mostly good enough 28

    Getting started with the community edition 29

Commercial editions: The ones you want 30

    Databricks on Amazon Web Services 32

    Azure Databricks 36

Summary 38

Chapter 4: Workspaces, Clusters, and Notebooks 39

Getting around in the UI 39

Clusters: Powering up the engines 42

Data: Getting access to the fuel 45

Notebooks: Where the work happens 46

Summary 49

Chapter 5: Getting Data into Databricks 51

Databricks File System 51

    Navigating the file system 52

    The FileStore, a portal to your data 55

Schemas, databases, and tables 55

    Hive Metastore 56

The many types of source files 57

    Going binary 58

    Alternative transportation 60

Importing from your computer 60

Getting data from the Web 62

    Working with the shell 62

Basic importing with Python	64
Getting data with SQL	66
Mounting a file system	67
Mounting example Amazon S3	67
Mounting example Microsoft Blog Storage	69
Getting rid of the mounts	70
How to get data out of Databricks	71
Summary	72
Chapter 6: Querying Data Using SQL	**75**
The Databricks flavor	75
Getting started	76
Picking up data	77
Filtering data	79
Joins and merges	82
Ordering data	84
Functions	85
Windowing functions	86
A view worth keeping	89
Hierarchical data	90
Creating data	91
Manipulating data	94
Delta Lake SQL	95
UPDATE, DELETE, and MERGE	96
Keeping Delta Lake in order	98
Transaction logs	99
Selecting metadata	99
Gathering statistics	101
Summary	102

TABLE OF CONTENTS

Chapter 7: The Power of Python 103

Python: The language of choice 103

A turbo-charged intro to Python 104

Finding the data 107

DataFrames: Where active data lives 108

Getting some data 110

Selecting data from DataFrames 114

Chaining combo commands 116

Working with multiple DataFrames 125

Slamming data together 132

Summary 136

Chapter 8: ETL and Advanced Data Wrangling 139

ETL: A recap 139

An overview of the Spark UI 140

Cleaning and transforming data 142

    Finding nulls 143

    Getting rid of nulls 144

    Filling nulls with values 146

    Removing duplicates 148

    Identifying and clearing out extreme values 150

    Taking care of columns 153

    Pivoting 154

    Explode 156

    When being lazy is good 156

    Caching data 158

    Data compression 160

    A short note about functions 163

    Lambda functions 164

Storing and shuffling data 165

    Save modes 165

Managed vs. unmanaged tables	167
Handling partitions	169
Summary	174

Chapter 9: Connecting to and from Databricks 177

Connecting to and from Databricks	177
Getting ODBC and JDBC up and running	178
Creating a token	179
Preparing the cluster	180
Let's create a test table	180
Setting up ODBC on Windows	181
Setting up ODBC on OS X	182
Connecting tools to Databricks	185
Microsoft Excel on Windows	185
Microsoft Power BI Desktop on Windows	186
Tableau on OS X	187
PyCharm (and more) via Databricks Connect	188
Using RStudio Server	191
Accessing external systems	193
A quick recap of libraries	194
Connecting to external systems	195
Azure SQL	195
Oracle	196
MongoDB	198
Summary	199

Chapter 10: Running in Production 201

General advice	201
Assume the worst	202
Write rerunnable code	202
Document in the code	202
Write clear, simple code	203
Print relevant stuff	204

TABLE OF CONTENTS

Jobs 204
    Scheduling 206
    Running notebooks from notebooks 206
    Widgets 208
    Running jobs with parameters 210

The command line interface 212
    Setting up the CLI 212
    Running CLI commands 213

Revisiting cost 220
Users, groups, and security options 220
    Users and groups 221
    Access Control 222
    The rest 225

Summary 225

Chapter 11: Bits and Pieces 227

MLlib 228
Frequent Pattern Growth 228
Creating some data 229
Preparing the data 230
Running the algorithm 231
Parsing the results 232
MLflow 233
    Running the code 233
Checking the results 236
Updating tables 236
Create the original table 237
Connect from Databricks 238
Pulling the delta 239
Verifying the formats 240
Update the table 241

A short note about Pandas	242
Koalas, Pandas for Spark	242
Playing around with Koalas	243
The future of Koalas	245
The art of presenting data	246
Preparing data	247
Using Matplotlib	248
Building and showing the dashboard	249
Adding a widget	249
Adding a graph	250
Schedule run	251
REST API and Databricks	251
What you can do	251
What you can't do	252
Getting ready for APIs	252
Example: Get cluster data	253
Example: Set up and execute a job	255
Example: Get the notebooks	258
All the APIs and what they do	259
Delta streaming	260
Running a stream	261
Checking and stopping the streams	264
Running it faster	265
Using checkpoints	266
Index	**269**

About the Author

Robert Ilijason is a 20-year veteran in the business intelligence segment. He has worked as a contractor for some of Europe's biggest companies and has conducted large-scale analytics projects within the areas of retail, telecom, banking, government, and more. He has seen his share of analytic trends come and go over the years, but unlike most of them, he strongly believes that Apache Spark in the cloud, especially with Databricks, is a game changer.

About the Technical Reviewer

Michela Fumagalli graduated from the Polytechnic University of Milan with an MSc in Electronics and TLC Engineering. She also got a master's degree in Big Data and Analytics, along with a Databricks Apache Spark certification.

She has studied and developed several machine learning models to help put data in the heart of businesses for data-driven decisions. Her other interests lie in the fields of reinforcement learning and deep reinforcement learning. After having gained experiences in various international companies, she now works for IKEA.

Introduction

So you wanna be a data analyst, data scientist, or data engineer? Good choice. The world needs more of us. Also, it is not only fun and rewarding work but also easy! At least if you're willing to put in the effort.

The bar for getting into heavy-duty data analytics has never been lower. You don't need servers, advanced Linux skills, or a ton of money to get started anymore. Graphical tools like Tableau and Power BI made small-scale data work available for the masses. Now Databricks does the same for huge datasets. Millions, billions, and trillions of rows are not a problem.

This book will ease you into both the tool and the field. We'll start by looking at the data analytics field in general – why it's hot, what has changed, and where Apache Spark and Databricks fit into the overall picture.

Then it's time for you to learn about how Databricks works. We'll spend a few chapters on learning how it works internally and how you actually use it – everything from getting around the user interface to spinning up clusters and importing data.

Once you know how to use the tool, it's time to start coding. You'll familiarize yourself with (Structured Query Language) SQL and Python, the two main languages for data analysis work. It doesn't stop there, we'll follow it up by digging deeper into advanced data wrangling techniques where we'll see a lot of the if's and but's you'll come across when you work with data in reality.

Finally, I'll drag you through a few short chapters with more advanced topics. Coming out of them, you'll have an understanding of how to run machine learning algorithms, manage delta loads, and run Databricks through the application programming interface (API).

And with that, you'll be ready to get started for real, solving small and large problems in the real world. This is an introductory book, but once you're through it, you'll have the tools you need to start exploring huge datasets around you or in your business.

Looking forward to seeing you around in the world of data experts.

CHAPTER 1

Introduction to Large-Scale Data Analytics

Let's start at the very top. This book is about large-scale data analytics. It'll teach you to take a dataset, load it into a database, scrub the data if necessary, analyze it, run algorithms on it, and finally present the discoveries you make.

We'll use a fairly new tool on the market called Databricks for all of this, simply because it's the most ambitious tool on the market right now. While there are other tools that can provide you with the same capabilities, more or less, no one does it like Databricks.

Databricks hands you the massive analysis power of Apache Spark in a very easy-to-use way. Ease of use is important. You should be spending time looking at data, not figuring out the nitty-gritty of configuration files, virtual machines, and network setups. It's far too easy to get stuck in tech. With Databricks, you won't.

But before we start, let's spend a few pages talking about what this data analysis thing is all about, what's happened the last few years, and what makes large-scale analytics different from the stuff you might have run in Excel or SQL Server Analysis Services.

Analytics, the hype

If you haven't lived under a rock the last few years, you've seen advanced analytics being referenced pretty much everywhere. Smart algorithms are winning elections, driving cars, and about to get us to Mars. They'll soon solve all our problems, possibly including singularity. The future is here! At least if you are to believe the press.

Yes, the hype is real, and a lot of people are getting very excited and ahead of themselves. Fortunately, while a lot of the talk might not be realistic in the near future and a lot of the stuff we read about really isn't that groundbreaking, there's a lot of useful stuff happening in the field – stuff that can help you make better decisions today.

Analytics really is one of the key drivers in pushing many industries forward. We probably won't have fully autonomous cyborgs around us for a while, but we can expect to get cheaper items with less issues being delivered faster, all thanks to smart analysts and clever algorithms.

A lot of this is already happening and has been for a while. Computer-based analytics has been around for ages and helped us get to where we are. Even hot topics like machine learning have been a staple of analysis work for a very long time.

Still, the world of analytics is changing. We have much more data now than ever before, computers are better equipped to deal with it, and – maybe most importantly – it's all much more accessible, thanks to Open Source Software (OSS) and the (also somewhat overhyped) cloud.

Analytics, the reality

So what is analytics really about? At its core, analytics is basically just a way to answer questions by finding patterns in data. It can be really simple stuff – most computer-based analytics in the world is probably still being done in Excel on laptops – or very hard, requiring custom software running on thousands of processor cores.

You do analysis in your head all the time. When you see dark clouds in the sky, you know it'll rain because similar clouds have in the past resulted in showers. If you drive the same way to work every day, it's likely you know with some certainty what the traffic will be like at any given point.

Basically, you've collected observations, or data points, and recognized a pattern with which you can make a prediction. Hopefully, you use those results to make smart choices, like bringing an umbrella or beating the traffic by leaving a bit earlier. Computer-based analysis in business is pretty much the same thing, just a bit more organized and based on more data points – (usually) collected in a structured way.

The results from companies doing analysis are all around us. Your insurance is based on analytics. So are your Netflix recommendations, your Facebook feed, and Goodreads suggestions. Google knows when you need to leave home to be at the meeting in time. Walmart can do a good prediction of your next shopping basket. The time it takes for the traffic light to switch from red to green is also based on analytics, even though it sometimes can feel as if those particular analysts aren't doing a good job (which they actually are – read *Traffic* by Tom Vanderbilt if you don't believe me).

Many a time, the analysis work that seems easiest is the hardest. The weather forecast you heard this morning is the result of some of the most intense analytics processing around. In spite of that, they aren't that much better than just assuming the weather today will be like yesterday. Keep that in mind – doing good analytics is hard, no matter what you hear or your gut tells you.

Most importantly for the sake of this book, analysis can help you and your company. There are a huge number of use cases within companies. The sales representatives want to know how to approach one type of customer differently to increase sales. Marketing wants to know which shade of gray will get the most eyeballs on the click campaign. The board needs to know if it makes most sense to establish presence in Sweden or Norway next. Store managers want to know on which shelf to put the milk. And of course many more. Analytics help you make guesses with higher accuracy.

Large-scale analytics for fun and profit

Large-scale analytics is, at its core, just more of the same. As the name implies, it's analytics, but on a larger scale than we've seen traditionally. There's a paradigm shift going on right now, and there's a very good reason for it.

There are three main things that have happened recently: a deluge of data even in normal companies (i.e., not just Facebook, Amazon, and Google), easy access to great tools, and a cheap way to access an almost infinite amount of processing power.

Traditionally, you had statisticians working with somewhat large datasets in SAS, SPSS, and Matlab on their own machine or on a local server or cluster. Underneath, you had a data warehouse that could offer up a huge amount of cleaned data from large database servers in privately owned data centers.

The same data warehouse was also used by business intelligence solutions. They made reporting and data discovery much easier. Basic analysis was made available to the masses, especially with the introduction of tools like Tableau and QlikView. This worked for a good long while. While it still works and will work going forward as well, it isn't enough. For new analytic needs, the methods of yesterday just don't cut it. As the data volumes increase and the business asks more advanced questions, new tools are necessary.

This is a natural progress. If you want to analyze thousands of rows, Excel is just fine. But as you move into millions of records, you probably move to a database instead. In the same way, you need to leave the classic tools behind if you want to run algorithmic operations on a very large dataset.

A single server will for instance have a hard time processing regressions for billions of combinations in a timely manner. Setting up a large data center with hundreds of servers doesn't make sense if you don't have the need for the machines to be spinning almost all the time. Most companies don't have that need.

You might, by the way, wonder what large is. Good question. As everything seems to grow rapidly, it's terrifying to put any value into print as it'll soon be outdated and laughed at. Just look at the company Teradata. When they were founded, in the late 1970s, a terabyte of data was an insanely large amount of data. Today, not so much. There are terabyte-sized drives the size of your thumbnail, and right now petabyte is the new big. Let's see how long Oracle's Exadata line of machinery keeps its name.

What is Big Data?

In reality, it's just a marketing term. It didn't start out that way, but over the years, there have been more definitions than it is possible to keep track of. Today, it's being used as a synonym for so many things that it's impossible to understand what the other party means without having them define it.

During the last few years, I've always asked people to do exactly that – to explain what they mean when they use the words. Here's a few things I've gotten back (in condensed form – I usually get a much longer explanation): "a lot of data" (whatever that means), "predictive analysis," "Hadoop," "analysis," "Twitter data," and "a lot of data coming in fast, in different formats" (based on the Gartner definition). This is just a small subset, but in every single case, there's another way to explain what they mean. My tip is don't use the term as the listening party probably won't hear what you think they hear.

It might be easier to think about it in terms of scalability. It turns large once you can't handle the data analysis without spreading the processing to many machines. That's when you need to look at new alternatives for handling the workload.

Machine learning is a good example of an application that works better the more data and processing power it gets. Running an advanced algorithm on a large set of data requires a massive number of CPU cycles. It could very well be that your desktop computer would never be able to finish it while a big cluster could do it in a matter of hours or even minutes.

So we've now established that data, software, and the cloud are the main drivers in the new world of analytics. Let's go through each of them and see what it is that really changed and how it affects data analytics.

Data: Fueling analytics

The most important driver in the new data analysis landscape is data. There is just so much of it right now. Creating more of it has never been easier or cheaper, and a lot of people are building stuff to produce more at an even faster pace. The growth numbers are ridiculous.

Consider the fact that every time you go to an ecommerce web page, tons of data are stored: what was shown to you, what you clicked, how much time you spent on the page, where your mouse pointer was, and much more.

Also know that everything you do/don't do also generates information. If you are recommended an article on *The New York Times* and you don't show any interest in it, you can be sure they make a note of it so their algorithms can try to do better next time. In total, your data trail of a single visit to a single page can easily produce many kilobytes or even megabytes of data.

This type of information flood is of course not limited to customer-facing aspects of the business either. Pretty much every part of every sector is enhanced to capture more and more data. While humans create a lot of it and applications even more, sensors are quickly taking the lead in producing digital bits.

While Internet of Things, IoT, is another term from the hype bucket, the number of sensors being produced and deployed is staggering. The already low prices keep falling; and the barrier to putting yet another sensor onto your watch, container, or truck is lowered every day. The car industry alone adds 22 billion sensors per year. Each and every one of them can create data thousands of times per second.

> **Is streaming the only way in the future?**
>
> It's getting more and more popular to stream data, passing data along as it comes in. While this is a perfectly valid way of handling data, especially in the context of web applications, it's not needed in all scenarios.
>
> How fast you need the data depends on how fast you need to make decisions. Online retailers, for instance, might change prices many times per second to adapt to the user visiting. Physical stores may only change their prices once a month. So it's the end result use that should decide how you should handle your data stream.
>
> That said, it's never a bad idea to stream data if you can. It's easier to set up a pipeline to let the data flow with minimal lag and process it in bulk, than the other way around. But don't sweat it if you get your data in bulk today. It's still how most data is being processed, and it'll be our focus in this book.

All these things add more data to the pile, to the tune of 1.7 megabytes per person, according to business intelligence company Domo. That is for every second, around the clock, every day of the year. I'll save you the trouble of adding up the numbers – it's 146 gigabytes per day. That's half a dozen of Blu-ray disks. Multiply that with 8 billion people, and you realize it's plenty. Quintillion might sound like a lot, but with the speed with which we increase data production, I wouldn't want to start a company called Quintilladata today. It might look a bit silly in 30-something years.

It's worth mentioning that there is a push against this data creation and collection from governments around the world, especially around personal data. There are several laws and regulations, like General Data Protection Regulation and China Internet Security Law, trying to stop companies from tracking individuals. The fear is a testament to how efficient the algorithms have become.

The most important thing from an analytics perspective is that there's a lot of interesting information in that data heap. Processing it in Excel on a $1,000 laptop won't really cut it. Neither will running it on an Oracle-based business intelligence platform or a SAS server cluster. To process very large amounts of data, you need tools that can handle it. Luckily, they exist. As a bonus, many of the best are free.

Free as in speech. And beer!

Most popular tools in the statistics and business intelligence world have been, and still are, expensive. They are aimed at large enterprise companies with the money to buy the technology they need. Unfortunately, that means that a lot of companies, not to mention individuals, can't afford the good stuff. This has been true not only for analytics software but for all types of programs.

Luckily, the Open Source Software, OSS for short, movement changed all that. Linux used to be a small oddity but has now replaced the Unix derivatives in most companies. The free Apache web server is the largest web server software; and its biggest competitor, Nginx, is also open source. There are many other examples as well. While the desktop client space is still largely dominated by Microsoft and Apple, the back end is moving more and more toward open source. Actually, even clients are relying more and more on OSS in the form of Google's Android, the biggest operating system in the world.

This trend is true within the analytics space as well. Most tools used today are free, both as in speech and beer. Most of the top tools used by analysts nowadays have OSS roots. At least partially. Not all of them though. Proprietary tools like Excel and Tableau are still popular.

The most important ones are open though, including languages like R and Python with thousands upon thousands of free libraries, which have democratized data discovery. You can download Python and a library like Open Computer Vision (OpenCV) through Anaconda for free and run it on your home computer.

Don't for a second think that the tools are just ok. These tools are the best in their field. If you look for instance at deep learning, all the top libraries are open source. There is no commercial alternative that beats TenSorFlow, Keras, and Theano. Free is not only good. It's the best.

There's also a different group of tools that help you handle the data you collect more efficiently, tools that help you slice the workload into smaller pieces. This is very important in large-scale analytics. The reason is that adding more power to a machine, scaling up or vertical scaling, has its limits and quickly becomes very expensive.

The alternative is to use a distributed architecture, also called scale-out or horizontal scaling. Instead of adding power to one server, you add more servers and have them work together. That requires software that can spread the workload across many machines.

The idea is to split the workload up into multiple parts that you can run in parallel. Then you can distribute the different units across many cores. If the number of cores you need is small enough, you can run this on a single machine. Otherwise, you run it on many machines.

The crux is that if you write code that only runs on a single core, you'll be limited by the top speed of that core. If you write code that can run in parallel, you'll get the combined power of all cores.

This is where you'll get into contact with products like Apache Hadoop/MapReduce and Apache Spark, both open source. You define what needs to be done in the correct way, and they help you distribute the work. They help you out along the way and make it as easy as possible for you to make the right decisions.

Impressively, they are free to download and use. You have access to the same toolset as anyone else in the world. No company, with the possible exception of the Silicon Valley and China giants, has tools better than what you can use. How cool is that?

If you have the inclination, its code is also freely available for you to look at. Even if you don't plan on coding anything yourself, it can sometimes be helpful to look at the code. The in-line documentation is in many cases very good.

There are however a few issues with Apache Spark, or any other tool that distributes data loads. Most importantly, they need to have underlying hardware to handle the processing. While prices of hardware are dropping, it's still expensive to build data centers and hire technicians, especially if you only need the power sometimes, like during a weekly load. Luckily, there's a solution for this issue. This brings us to the third component in our paradigm-shifting troika.

Into the clouds

In the 1940s, Thomas Watson of IBM famously claimed that the world market for computers was about five. Although he was somewhat off mark, you could today exchange his use of "computers" to "computing power providers" and not be too far off as a handful of companies are handling more and more of the world's processing needs.

A large amount of the computing power is still located on-premise at companies. Most companies with more than a few hundred people employed have some kind of data center that they own. But the number is quickly dwindling. More and more companies, especially new ones, are instead handing the technology infrastructure part to someone else.

Cloud computing as it's called, a term popularized by Google's Eric Schmidt, has become the new norm. Instead of running your own technology stack, you just buy what you need when you need it. It's kind of like water or electricity. Most companies don't have their own utility plants. The same should be true for computing power.

The analogy only works so far as your data is probably less fungible than the electrons flowing through your socket. Still, having access to a near infinite amount of processing power while at the same time not having to employ even one person or build a single rack is hugely important to smaller companies. It levels the playing field quite a bit and is one reason smaller startups so easily can scale.

Together, cloud computing and distributed computing have opened up an amazing opportunity for analytics-minded people. Problems that couldn't easily be solved in the past or would take far too long time to solve can suddenly be processed even by minor companies.

Larger companies benefit as well. Workloads that formerly took days to run can suddenly be run daily, hourly, or in some cases instantly. Plus, no matter how big you are, not having to build out another data center is always a good thing.

Best of all, the price will often be lower as you'll ideally only pay for what you use. A server in your own data center will cost you money no matter if you use it or not. In the cloud, you only pay for what you use, so you can spin up 100 servers for an hour and then let someone else use them instead.

So that's our three core changes: data, tools, and cloud computing. One problem remains though. Even though things have gotten easier, it's still too complicated for many to set everything up from scratch. Configuring infrastructure on Microsoft Azure or Amazon Web Services (AWS) is easy. Doing it wrong is even easier. Setting up Hadoop Core or Spark requires a lot of reading and testing by a tech-minded person. If only there was a better way...

Databricks: Analytics for the lazy ones

That's where Databricks comes in. Yes, finally we're here. Ability to handle massive data amounts? Yes. Based on best-in-class tools? Check. Cloud powered? Exclusively. All this converges in a simple-to-use tool that practically anyone can start using within a matter of minutes. You get the power of Apache Spark in the cloud, with an almost infinite amount of storage space, available to anyone, for a fraction of the price it would cost to set it up in a data center.

This is the true power of the new times. You don't have to think as much about the underlying hardware and software anymore. If you want to optimize things, it's still good to know how stuff works, which is why we'll go through it later in the book. If you don't want to, you'll be pretty well off anyways. Instead of thinking about infrastructure, you can focus on business problems and how you can solve them.

Even better, you also get help with running the software. Clusters are scaled automatically and shut down if you don't use them. That will save you a lot of money and make your life easier. Everything is essentially taken care of for you. You can focus on analyzing data, which is what you should be doing.

These types of Software as a Service (SaaS) tools are on the rise within the analytics field. More and more companies want to outsource the backend stuff and instead have their expensive data scientists and data engineers focus on the business task at hand.

Databricks and tools like DataRobot, Snowflake, and Looker are all part of this new trend. Older players are trying to catch up, but so far they haven't made much of an impact. While that might change, it's likely that many of the new players are here to stay. Databricks stands a better chance than most. They are available at the two biggest cloud platforms, Azure and AWS, have a reasonable pricing model, and – maybe most importantly – are run by the initial developers of Apache Spark. They have the foundation for a very successful future, and with the help of this book, hopefully so do you.

How to analyze data

Before we dive into how you use Databricks, let's consider what you can actually do when you have all of this set up. The answer is, too much. The opportunities are almost endless. This can in itself be a limiting factor. When you can do almost anything, it can be hard to actually do anything and harder still to do the right thing.

It's common to try and find something of interest by looking aimlessly at outliers and odd patterns, frequently referenced as exploratory analytics. This is sometimes a good way to find interesting anomalies and indications of issues that you can then pursue in a larger scale. It is, however, rarely the most efficient way of creating business value. Try instead to always start with the problem.

CHAPTER 1 INTRODUCTION TO LARGE-SCALE DATA ANALYTICS

In his book *Thinking with Data*, Max Shron provides a very good process for working with data analytics projects. While it's not the only process, I've found it to be very helpful both for setting up my own projects and evaluating others. There are four basic steps you need to consider: context, need, vision, and outcome.

In context, you define what you are trying to achieve and what stakeholders are interested in the problem you will hopefully solve. You also define if there are higher-level goals you need to consider.

Need is where you clearly need to state what the project, if successful, will bring to the table. What will you be able to do that you just couldn't do earlier? This is crucial to understand the problem at a practical level.

You also need to define what the result will look like. This is the vision. It can be a mock sketch or a handful of made-up rows with the expected types of values you'll get. Also define the solution logic.

Outcome is probably one of the most important parts, one that is frequently overlooked. How will the result be used? And by whom? Who will own the resulting solution? It's far too common to see good work being dumped because there's no real way to integrate it with the current business process.

If you can come out of the preparation work with all questions clearly answered, a very large part of the battle is already done.

A small note: Although this book is about large-scale analytics and that's what I've mostly been talking about here, a lot of data is not always necessary. Sometimes small datasets can be used to highlight important things.

But if you want to do good predictions about the future using data, then more is usually better. If you know the weather trends for the last 100 years, you'll probably be more accurate in your forecast than if you only have them for the last 5 years. Then again, considering the current weather trends, you maybe won't...

Still, even if more data isn't needed, it's good to have it around. It's much easier to filter out information you don't want or need than to create or gather a new one. In many cases, it's not even possible.

In many cases, you'll realize that the data you need to finalize your work just doesn't exist yet. Either it's not currently being captured or the information you need just isn't available in your company or system. If so, you need to get that information. How to do that is outside the scope of the book, but I can recommend *How to Measure Anything: Finding the Value of Intangibles in Business* by Douglas W. Hubbard.

CHAPTER 1   INTRODUCTION TO LARGE-SCALE DATA ANALYTICS

Large-scale examples from the real world

So let's look at what companies are doing with analytics. The list could be almost endless as pretty much every mid- or large-sized company in the world is doing some type of analytics. But to give you a flavor of what is possible with large-scale data analytics, here are three examples.

Telematics at Volvo Trucks

Volvo Trucks and Mack Trucks are using thousands of sensors on their trucks to collect status data. This information is fed into a solution that tries to predict if the truck is about to fail. Instead of the driver being stranded, he's early on advised to drive to a service station and have the problem fixed before it happens.

It's interesting to note that it's not just information about the actual truck that's relevant. They also collect information about the surroundings to see if things like air pressure or altitude are important factors.

The result is less breakdowns and standstills, which is an important cost driver for most delivery companies. This solution is only a small piece of all the analytics that goes in to transportation logistics. It's a field that has used large-scale analytics for more or less as long as computers have been readily available.

Fraud detection at Visa

The financial sector has also used large-scale analytics for many years, especially the credit card companies. They have a keen interest in quickly figuring out if a transaction is fraudulent or not.

Visa, for instance, is running predictive analytics in real time for all purchases being made using their cards. The algorithm is looking for patterns that are similar to earlier fraudulent ones. If it seems fishy, the purchase is declined and the card holder informed.

During the last decade or so, fraud has declined dramatically, thanks to these methods. While the volume has increased by more than 1,000 percent, the fraudulent ones have declined by two-thirds. And the system is getting better by the day, in both directions.

You might actually have had an experience of a false positive, a fraud warning for a correct transaction in the past. That can happen if you do an unusual series of purchases that is out of the ordinary for you, especially online. If so, it probably was a while since it happened. Visa and Mastercard are pretty good at what they do today and can better understand patterns and understand if your unusual purchase is ok.

Customer analytics at Target

The next example, which might be apocryphal, is quite old now but still very relevant. Almost 10 years ago, retail giant Target sent a letter to a young girl, congratulating her on being pregnant. Believing this to be false, the father of the family got really angry only to later have to apologize as it proved to be true.

Target figured this out based on the girl's buying habits. Her sudden change in behavior put her in a new cluster, and the algorithm quickly realized that she was expecting, apparently before her family.

This example could absolutely be true. Most retailers have a very good understanding of you based on your purchase history. This is especially true for very large retailers that can compare you to millions of other customers and even more true for ecommerce sites.

This also highlights a growing fear of integrity intrusion. Customers don't like that companies know so much about them. But the thing is it's all in the aggregate. It's not like Google is logging in to your email account and checking to see what you sent to your friends.

Targeted ads at Cambridge Analytica

Well, actually, let's not use this as an example, not a positive one at least. Still, keep it in mind when you do your analysis. Don't forget ethics when you are digging around in the data piles you find. There is no Hippocratic Oath for analysts, but sometimes it feels like there should be one. Be nice, predict for good.

Summary

With this chapter behind us, I hope you have an intuition about where large-scale data analytics fits in the bigger picture and why Databricks has become so popular. While analytics has been around for a long time, increases in data volumes, easy access to cloud processing power, and great open source tools have enabled new opportunities in the analytics field.

One of the most popular frameworks available, Apache Spark, has risen to the top on the tool side. Databricks has made using it in the cloud easy. With a few button clicks, you can have it handle billions upon billions of rows of data without even breaking a sweat. Throw petabytes at it and it'll just shrug.

It's also a complete toolset with everything you need to load, scrub, and analyze data using a huge amount of resources that you can rent from pay-as-you-go cloud providers such as Microsoft or Amazon.

Databricks probably won't replace your data warehouse nor your desktop analytics tools even though it could take on those chores. It's the power tool you pick up when everything else fails – the beast to unleash when things get really hairy.

Still, the tool doesn't do all the work. Ultimately, you need to consider what to do, how it will help your business, what the result will look like, and who will take your findings and make sure they are implemented and maintained.

Next, let's look at what Spark is, how it works, and what makes Databricks special.

CHAPTER 2

Spark and Databricks

Apache Spark has taken the world by storm the last few years. In this chapter, we'll go through why it's so and what Apache Spark actually is. We'll also take a look at what Databricks brings to the table.

To get a better understanding of the tool, we'll take a look at the architecture. This will help us understand how jobs are run in the system. While not necessary, it'll help us solve problems in the future.

We'll also take a closer look at data structures in Apache Spark. While we'll go into Databricks File System (DBFS) in more depth in the next chapter, we'll take a quick look at storage already here.

Finally, we'll take a quick look at the components that are offered on top of Apache Spark. They are not necessary, but they might be exactly what you need to get your use case done. A couple of them we'll come back to in later chapters.

A quick note: Both Spark and Databricks are changing fast. Most code in this book was run on Spark 2.4.4. While most of the things here will be true for earlier and later versions, just keep in mind that things might have changed.

Apache Spark, the short overview

Apache Spark is a tool for processing large amounts of data. It does this efficiently across a large number of machines, or nodes. While you can use Apache Spark for a number of different use cases, it's mostly being used for analytics.

The biggest reason people like Apache Spark is the speed. It is fast when used with large datasets. While sizing is a bit vague, it basically works well where single-server systems just can't cope anymore.

Compared to traditional solutions, and even many other cluster or parallel processing solutions, it can be up to several times as fast – up to 100 times faster in some cases. This is especially true when compared to the other popular open source alternative, Hadoop/MapReduce.

If you're an IT veteran, feel free to roll your eyes now. Claims of betterness are a disease in the business. Everyone can somehow always prove they're much faster than the competition. Still, properly used Apache Spark really is a well-written piece of software that can produce impressive results not only in benchmarks but in reality.

It is also usable with huge amounts of data and machines. The core architecture makes Apache Spark very scalable. It can run on many thousands of nodes if you need it to. Or you can run it on just one computer, although that's almost always a bad idea unless you're testing it out.

Scalability is always good, but even more so when you're running jobs in the cloud. When you need power, just add more nodes. Once you're done with them, shut them down. If your jobs are linear, your wallet decides how fast you get your results.

As I'm writing this, I have a nine-node cluster on Azure running in the background. Each node has 64 cores and 432 gigabytes of working memory. While it's not supercomputer-beating numbers, it's enough for many use cases. If a colleague wants to run another test case, they can spin up their own, separate cluster.

It's also cheaper to run many smaller computers compared to buying one large machine. In the preceding example, I'll pay a handful of euros or dollars to get the result I need. Then I'll shut everything down and pay nothing until I need it again.

Another great thing is that Apache Spark is easy to use once installed. You can use a large number of different programming languages to communicate with it. Most important in our case is that it handles both Python and R well.

It also offers several libraries to make your life easier. The SQL and DataFrame ones are something we'll use a lot in this book, for instance. If you have prior experience with databases, these will ease you into Apache Spark.

Furthermore, you don't need to use just one, proprietary file format. Apache Spark works well with a large number of file types. We'll talk about this in a later chapter, but you can use Apache Parquet, Optimized Row Columnar (ORC), JavaScript Object Notation (JSON), or a number of other options if you want to.

The good thing about this is that you easily read data you already have available. Even if your data is unstructured, like images or sound files, you can read it into Apache Spark and process it.

It is also open source, meaning that you can actually download and compile it if you want to. If you are a really good programmer, you might even add a few features to it. Even if you can't or don't want to, it's possible to peek inside and see what actually happens when you run a function call. The Spark code is actually well documented. If you feel so inclined, head over to `https://github.com/apache/spark` and look at it.

Since 2013, Spark has been maintained by the Apache Software Foundation. If you're not familiar with Apache Software Foundation, it's a nonprofit powerhouse of a company. They maintain many of the best open source tools available on the market today. Many large commercial companies, like Facebook and Google, hand over their tools to Apache Software Foundation to secure their maintenance. Among the tools, you'll find Apache Hadoop, Apache Cassandra, or Apache HBase.

All of this has made Apache Spark the go-to tool for large-scale analytics. It's used by tens of thousands of companies with large-scale analytics needs. Fortune 500 companies use it, as do much smaller businesses. More and more of them are turning to Databricks for their Apache Spark needs.

Databricks: Managed Spark

Databricks is Apache Spark packaged into a cloud service – fully managed and with a bunch of goodies on top. Being built by the same people who created Spark in 2009, it uses the core engine in the best of ways. Most importantly, it removes all the techy parts of getting started.

While Apache Spark is somewhat easy to set up for a technical person, it still requires you to understand network and operating system technology. With Databricks, you only need to click a few buttons. Everything else happens under the surface.

This means you can focus on your data challenges instead of digging through configuration files. You don't have to care about architecture, setup, or much anything else related to the technical stuff of Apache Spark. Everything is being taken care of for you. That said, knowing what goes on under the hood is beneficial.

Databricks is also cloud optimized and is good at scaling both up and down. If you don't run a constant load 24/7, this is great. You define the perimeter at start, and then Databricks does its magic in the background. That way, you have power when you need it and no cost when you don't.

This pay-as-you-go model is a big benefit of Databricks for most companies. You can run your monthly job with a huge number of nodes and then shut them down. Compared to running local servers, the savings can be massive.

You also get an interface that is actually good. It's minimal, easy to use, and built for collaboration. You and your colleagues can work in the same notebook at the same time, for instance. That makes co-development possible even over distance.

While working with the built-in notebooks is the most convenient way to interact with Databricks, you can use other tools as well. You just connect using one of many connectors available. You can even use desktop analytics tools like Microsoft Power BI, which works like a dream.

Since Databricks runs Apache Spark, you can use it to extend your on-premise setup. This can be good if you have large, infrequent jobs. It's not the same user interface experience, but you might avoid buying more servers for a once-a-quarter job.

The far side of the Databricks moon

While there are a lot of positive things to say about Databricks, some problems exist. You shouldn't throw yourself into the cloud-only analytics space without considering the downsides. It isn't for everyone.

One drawback is that Databricks is only available in the cloud. While that is in vogue right now and an excellent use case for clustered systems, there are still many companies who want to run stuff on-premise.

Others might want to use cloud providers other than Amazon and Microsoft. Right now, you can't get Databricks on Google Cloud Platform or on the smaller players in the field, like IBM and Oracle. While this is probably not a problem for most, it's still a limiting factor.

Another issue with the platform is that most of it is hidden. That's great while everything is working but less so when things go wrong. A lot of logs are available, but it can still be a pain trying to figure out why a cluster suddenly just decided to crash. This will happen, by the way.

Finally, you need to consider pricing. Apache Spark is free. Databricks, with all features enabled, is far from it. Running large jobs, you can easily run into a five-figure number of dollars per month for a single cluster in a single workspace – or even faster if you go nuts and aren't careful. I've seen examples of developers running Cartesian products by mistake on large datasets on a big cluster overnight. That quickly gets expensive.

Overall, Databricks is for most a better solution than trying to roll your own Apache Spark installation. It's just so darn convenient to spin up ten nodes, run a large job, and then shut everything down. The future also looks bright with more and more companies signing up. Even Google Trends shows a clear trend upward.

Spark architecture

Now you hopefully have an understanding of what Spark is and why it's being used. Let's dig a bit deeper and look at the nuts and bolts of the system. While you strictly speaking don't need to know about this to work with Databricks, it'll help you in your daily work.

The first thing to understand is clustering. A cluster is basically a number of machines, commonly referred to as nodes, working together as one unit. The reason you use clusters is to get more done in a shorter amount of time while not spending too much money.

There are large servers, like the IBM P-series, with hundreds of cores. Usually, they cost a lot more than a number of smaller machines with the same power combined. There is also a limit to how much you can cram into one physical box, while clusters can scale over many machines – thousands upon thousands if necessary.

Clusters can also be built to be more resilient. If your server crashes, everything stops. If a few nodes disappear in a cluster, it won't matter much if you have nodes remaining. The workload will just be handled by the remaining ones.

I should mention that while Apache Spark runs in a cluster, it technically doesn't have to be a multi-machine cluster, but on a single machine Spark can't really stretch its legs. It also doesn't make much sense to run any type of overhead on a single machine, so only use this for testing.

The cluster concept is not new to Apache Spark and really not all that new within the field of computer science. Like many other things in IT, it's been around since at least the 1960s. Historically, it's mainly been used for large-scale installations. High-speed networks and data amounts increasing faster than processing ability have made the technology in vogue for more use cases.

> **WHAT ABOUT MASSIVELY PARALLEL PROCESSING?**
>
> Another technology term you will probably come across is Massively Parallel Processing, or MPP. You will mainly see this in connection to commercial products like Teradata, Greenplum, and IBM Netezza.
>
> The core concepts are similar to Spark's clustering, and they are looking to achieve the same thing. The main difference is that MPP systems are more tightly coupled. Nodes in computing clusters are more independent.

MPP solutions frequently bundle together software with hardware for this reason. By building more connected systems – sometimes with proprietary hardware components – they can achieve great performance, usually at a high price.

That said, there is no clearly defined definition. For that reason, you might hear people use the terms synonymously. Apache Impala, for instance, uses the MPP term even though it's running on top of a normal cluster just like Apache Spark.

Luckily, if you're not buying hardware, it doesn't really matter much. So for cloud solutions, you can pretty much just forget about it. Just do a price/performance comparison between the solutions you are considering.

Apache Spark processing

Apache Spark is a software that runs on top of the cluster. It basically handles the jobs for you. It looks at what needs to be done, splits it up into manageable tasks, and makes sure that it gets done.

The core architecture is master/slave. The driver node is the controller and the head in the setup. This is where the status and state of everything in the cluster is being kept. It's also from where you send your tasks for execution.

When the job is being sent to the driver, a SparkContext is being initiated. This is the way your code communicates with Spark. The driver program then looks at the job, slices it up into tasks, creates a plan (called a Directed Acyclic Graph [DAG] – more about that later), and figures out what resources it needs.

The driver then checks with the cluster manager for these resources. When possible, the cluster manager starts up the needed executor processes on the worker nodes. It tells the driver about them.

The tasks are then being sent from the driver directly to the executors. They do what they are asked for and deliver the results. They communicate back to the driver so that it knows the status of the full job.

Once all of the job is done, all the executors are freed up by the cluster manager. They will then be available for the next job to be handled. That's pretty much the whole flow and all components.

Does all this seem a bit complicated? It really isn't. Still, if it makes you uneasy just to think about all these components, the good thing is you don't have to care. In Databricks, you just tell the system how many workers you need and how powerful they need to be. The rest is done for you. Virtual machines will be spun up, Linux images with Spark applied, and in a few minutes you'll be ready to go.

Working with data

With processing handled, let's turn our attention to data. The core data structure in Apache Spark is called the Resilient Distributed Dataset, or RDD. As you'll see in a bit, this is probably not the level you'll be working on. Still, it's what Spark uses underneath the surface.

As the name implies, the data is spread across a cluster of machines. These parts are called partitions and contain a subset of data. You can define the split limits yourself (e.g., to make daily partitions) or let Spark do it for you.

RDDs are also resilient. This has to do with how Spark works. RDDs are immutable which means they can't be changed. So when changes are done, new RDDs have to be created. Frequently, there'll be a lot of them in a job. If any of them are lost, Spark can recreate them by looking at how they were created in the first place. Sounds strange? It'll be clear in a bit.

Note that traditional Apache Spark does not replicate data. This is for the storage layer to handle. For instance, on Azure this is being handled by the blob storage. So Databricks doesn't have to think about it.

The final word is datasets, basically a chunk of data. In RDD the data is schemaless. This is great if you have unstructured data or your use case doesn't match with traditional formats. But if you have organized information, you probably want to work with something a little bit more adapted.

For this reason, there are other ways to work with data in Apache Spark: DataFrames and datasets. Unless you have good reason to use RDD directly, you should consider these two instead. It's both easier and better optimized.

DataFrames are like tables in a database. You can define a schema and make sure the data adheres to it. This is what most use cases in businesses need. It's also what we'll use for almost all of this book. By the way, don't confuse the Spark DataFrame with Pandas or R dataframes. They are similar but not the same.

Datasets are pretty much DataFrames, but strongly typed. This makes it possible for the compiler to find data type issues early on. While this is helpful, it only works in languages that support it. Python and R do not. So you won't be seeing it in this book. If you happen to be a Java enthusiast though, you should use datasets.

So we'll use DataFrames for most of our coding. For Apache Spark, that doesn't matter. As I mentioned in the beginning of the chapter, it'll use the same code as for RDD internally. So let's see how that happens.

Data processing

On a high level, RDDs support two things: transformations and actions. Transformations are operations on RDDs that return one or more new RDDs. As you might recall, all RDDs are immutable which is why new RDDs have to be made.

If you build up a chain of transformations, maybe combining sorting and grouping, nothing will happen immediately. This is because Spark does lazy evaluation. It basically just builds a map of everything that you asked for in an optimized order. This is the Directed Acyclic Graph, or DAG, we talked about earlier. Let's look at an example in Python:

```
df = spark.sql('select * from sales')
df.select('country', 'sales')
    .filter(df['region'] == 'EU')
    .groupBy('country')
```

Don't worry if the code looks confusing, you'll get familiar with it in Chapter 7. It picks up data from a sales table and then groups the data to sum sales per country. A filter makes sure we only use data from EU countries.

No matter if you have ten rows or billions of rows, running this will take a split second. The reason is that we only do transformations. Spark realizes this and decides not to bother the executors with it just yet. Let's look at another statement:

```
df2 = df.select('country', 'sales')
    .filter(df['region'] == 'EU')
    .groupBy('country')
```

We now assign the result to a new DataFrame. In this case, you might think that the assignment will create an action, but it won't. It's not until you actually do something with the DataFrames that you will end up creating an action that triggers work. Let's show a couple of actions:

```
display(df2)
df.count()
df.write.parquet('/tmp/p.parquet')
```

Actions are what get work going. They return non-RDD results, like data to the driver or to storage. In the preceding example, the first one runs a collect to pick up rows. The second one shows a count of rows. The third saves to a file system. It makes sense if you think about it. All of these actions require results to be available, which you only have once the processing is done.

These manipulations and actions are what the driver needs to consider when you pass on a job. The lazy evaluation makes it possible for it to optimize the whole job instead of just one step.

In a later chapter, we'll actually look at DAGs to see how they work more in detail. Looking at them can be helpful if you think that execution takes longer than expected. You can usually find hints of issues in the execution plan.

Storing data

There's also a storage layer to consider. A cluster needs data to be available on a shared file system. In the world of Apache Spark, you'll frequently see the Hadoop Distributed File System, HDFS, being used. This is not a requirement though.

If you store data in Databricks, you'll end up using the Databricks File System or DBFS for short. It's a distributed file system that you can access through the notebooks, code, and the command line interface (CLI).

Note that you don't have to store data within Databricks. You can use something like Azure Blob Storage to store your data and just connect to it. This makes sense if you plan on using multiple tools or workspaces.

The default format if you create a table with no options is Parquet with snappy compression. It's a pretty good choice for many analytics use cases. Sometimes other formats will do the work better. We'll get back to this discussion when we look at Databricks in more detail.

Tables you create in Databricks will end up at the /user/hive/warehouse path in DBFS. You can go there to look at all the databases, tables, and files. There are better ways to look at the structures in the database than using the file system. Sometimes, however, you need to work with the files directly, especially to clean up failed operations.

Data is being picked up from DBFS or whatever file system is being used. Work is being done one partition at a time per node. This is important as data spread can be crucial to your performance, especially if the data is skewed.

Cool components on top of Core

Most stuff we've talked about so far is related to Apache Spark Core. There are a number of components you can use on top of it to make your life even better. While they're a part of the Apache Spark ecosystem, you can freely ignore them and use other tools for the same job instead. The benefit of using the Spark ones is that they run on top of the Apache Spark engine. This means they are optimized for working well in parallel environments.

The four components are Spark Streaming, Spark Machine Learning library (MLlib), Spark GraphX, and finally Spark SQL.

Spark Streaming is used for handling continuous data integrations. Imagine for instance that you want to process tweets as soon as they come in. This requires a different mind-set from how you'd work with batches. With this library, you get a high-throughput, fault-tolerant solution out of the box.

Spark MLlib is a library containing a number of algorithms for machine learning tasks. It's optimized for Spark and will use the full power of the cluster setup. The biggest drawback is that it contains nowhere near as many options as scikit-learn on Python or CRAN on R. If you find what you need however, MLlib will work well with Spark.

As the name implies, Spark GraphX is used for handling graph data. If you have a use case connected to social networks, logistics, or something similar, GraphX can help you. Like with the other libraries, there are other options out there. What you get by using this one is the speed and performance Spark can bring. In many cases, it can be hard to make other packages scale well.

Lastly, we have Spark SQL. This one we'll use a lot in this book. It's a module that makes it possible for you to run SQL queries on Apache Spark as you would on any relational database. This makes it much easier to quickly work with data and lowers the barrier of entry for traditional analysts.

Summary

Over the last few pages, we've gone through what both Apache Spark and Databricks are. You should now have an understanding of what it is that makes the tools stick out and where they fit into the data analytics ecosystem.

We also talked about how data is being processed in the tool, with drivers handing tasks over to executors. You've seen how Apache Spark builds on top of cluster technology and how that makes it fast.

Then we went through data both from an internal and external perspective. We read up on RDDs, DataFrames, and datasets and how lazy evaluations hold transformations back until there's an action.

To round everything off, we quickly went through the four packages that are being offered on top of Apache Spark. One of them, Spark SQL, you'll be using soon. The others will come back in the latter part of the book.

Now that you have an understanding of how Apache Spark works in general, it's time to start digging into Databricks. To do that, you need to get up and running with the tool. That's exactly what we'll do in the next chapter.

So next up, working with Databricks.

CHAPTER 3

Getting Started with Databricks

In this chapter, I'll show you how to get Databricks up and running in different cloud solutions. We'll go through the different options you have to choose from and the pricing model that comes with them.

By the end of the chapter, you'll have a working installation on either Amazon Web Services or Microsoft Azure. With that, we'll be in good shape to move into data loading and development in the next two chapters.

Cloud-only

Unlike most traditional software, Databricks can't be downloaded and installed on your laptop or desktop computer. There's not even a version available for your company servers. The only way you can use Databricks is in the cloud. It's not cloud-first, it's cloud-only.

While this is a bit limiting to large enterprise companies, it makes things much easier for the company. Instead of updating their software every couple of years, they can actually offer you updates all the time. Right now, you get new stuff on a monthly basis, but Databricks could basically do this on an even faster pace. If you think that it isn't great to develop stuff in a changing environment, don't worry. There are stable versions you can choose for important production jobs.

Unfortunately, Databricks is not available on all cloud platforms. They started out on Amazon Web Services and have since then only expanded to Microsoft Azure. Luckily, these two are the biggest players in the space, so it's likely you'll have at least one of the two in your company. Also, it's likely they'll show up on Google Cloud Platform and other clouds eventually.

Community edition: No money? No problem

Even if you don't have access to commercial cloud through your employer and don't have a lot of money to spend on cloud resources, you can still use Databricks. They have been nice enough to offer up a free version that you can use to try the software out. Both software and hardware are free – no credit card information required.

This version is of course a bit limited and mainly intended for you to play around with the system. You won't be crunching terabytes of data on the community edition. What you will get is a feel for the user interface and the notebooks.

With the free version, you get a working system with just a 6-gigabyte driver node. No worker nodes, so there's no real distribution of work happening here. There's no collaboration available on the notebooks, and only three users can connect to it. It's also a public environment.

While this is not something you can actually use for any serious work, it's still great that you have the option. In my opinion, all software providers should offer an easy way for you to easily test out the product before you spend any money on it. So kudos to Databricks for doing this and even covering the hardware costs.

Mostly good enough

Most of the examples in this book can be run on the free version of Databricks. I've tried to use smaller datasets so you can actually get everything working. That said, you won't see all the options available in the top version; and some core features, like jobs, aren't available at all. Still, the base version is more than good enough to get started and learn the fundamentals of Databricks.

If you want to try the real thing, Databricks is offering a 14-day free trial. That's enough for you to get a feel for all the features and see what you get for your money. Consider this option once you've gone through the book and have a couple of weeks to really dig into Apache Spark with Databricks. Note that this trial only covers the Databricks software; you still need to pay your cloud provider for the infrastructure. That said, if you are new to the cloud provider, there's often some kind of free credits to newcomers. Keep an eye open for those deals.

CHAPTER 3 GETTING STARTED WITH DATABRICKS

Getting started with the community edition

Go to databricks.com. Up in the corner, you'll find TRY DATABRICKS. Click the button, and you'll end up on a general page nudging you to go with the free trial. Ignore that for now and click the right GET STARTED button (Figure 3-1).

DATABRICKS PLATFORM – FREE TRIAL	COMMUNITY EDITION
For businesses looking for a zero-management cloud platform built around Apache Spark	For students and educational institutions just getting started with Apache Spark
<ul><li>Unlimited clusters that can scale to any size</li><li>Job scheduler to execute jobs for production pipelines</li><li>Fully interactive notebook with collaboration, dashboards, REST APIs</li><li>Advanced security, role-based access controls, and audit logs</li><li>Single Sign On support</li><li>Integration with BI tools such as Tableau, Qlik, and Looker</li><li>14-day full feature trial (excludes cloud charges)</li></ul>	<ul><li>Single cluster limited to 6GB and no worker nodes</li><li>Basic notebook without collaboration</li><li>Limited to 3 max users</li><li>Public environment to share your work</li></ul>
GET STARTED	GET STARTED

Figure 3-1. *The Community Edition is a good way to get familiar with Databricks without spending any money*

Fill in the information they ask for and verify that you aren't a robot (Figure 3-2). There's a checkbox in the end as well. Check that one if you want some information about Databricks. I have actually signed up myself and don't get many mails, so they won't spam you. Feel free to sign up to keep updated on what's happening with the company.

Make sure you're using an email address you can access. Wait for a minute or two and then check your email and look for an email from Databricks. In the mail, you'll get a link that you need to click to verify that your address is correct. Do so now.

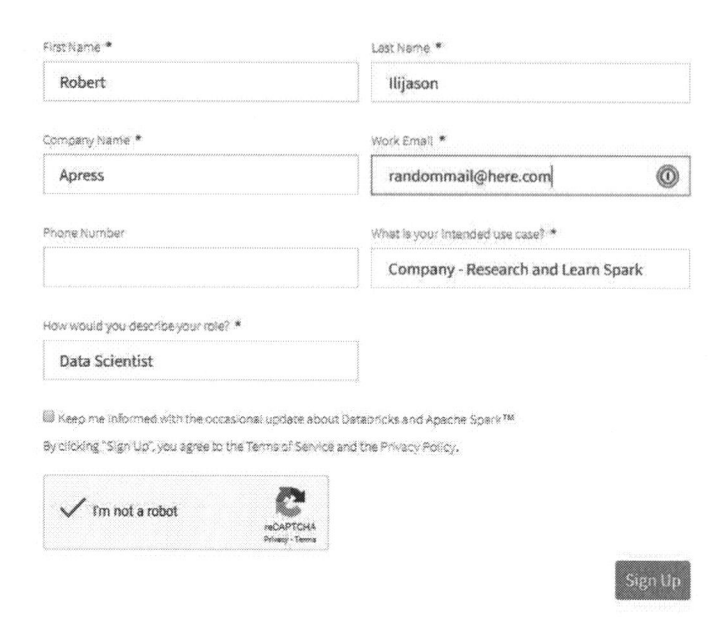

Figure 3-2. Don't forget to check the spam box if you don't get a mail within a few minutes

You'll now end up on a page where you need to create a password. Like on many other services, you need to create an unnecessarily complicated one. Make sure you have a capital letter, a number, and a special character in there. Once you have this filled out, you'll get dropped into your new workspace. Yep, you are ready to go.

Commercial editions: The ones you want

If you want the real Databricks, you'll have to pay up for it. While the community edition is good, it won't help you solve anything you couldn't do on a single machine before. To get all of the power, you need to go with one of the commercial editions.

There currently are two options available, Amazon Web Services and Microsoft Azure. The difference between the options is very slim. If you already use one of the providers, keep using that one. Otherwise, you can more or less flip a coin to decide which way to go, if using Databricks is the only thing you need to consider. If you're new to the cloud, you might want to lean toward Microsoft as they are easier to get started with, as you'll see later in this chapter.

No matter what cloud option you pick, you have three main feature sets to choose from. They are, from the most basic to the most advanced, Data Engineering Light, Data Engineering, and Data Analytics (Figure 3-3).

Data Engineering Light will give you the core features with fixed sized clusters and notebooks. You'll also have the option of running jobs with monitoring. That's pretty much it. This is basically a lightweight version for running well-known jobs in production.

Next up is Data Engineering. This adds quite a lot. You'll be able to schedule jobs via notebooks, which means you can do workflows easily. Clusters will be able to scale and shut down automatically, and you get access to machine learning stuff and Delta Lake. Pick this if you want a lot of features, work mostly on your own, and mainly work with putting stuff into production.

The highest tier is called Data Analytics and mainly adds collaborative stuff. All the extra features are related to working together in a team and integrating external tools, such as RStudio. So if you work within a team of data analysts, this is probably the way to go.

	Data Engineering Light	Data Engineering	Data Analytics
Managed Apache Spark	✓	✓	✓
Job scheduling with libraries	✓	✓	✓
Job scheduling with Notebooks		✓	✓
Autopilot clusters		✓	✓
Databricks Runtime for ML		✓	✓
Managed MLflow		✓	✓
Managed Delta Lake		✓	✓
Interactive clusters			✓
Notebooks and collaboration			✓
Ecosystem integrations			✓

Figure 3-3. *Using jobs is a good way to lower the cost*

On top of the base packages, you have a bunch of extra options around security that are cloud provider specific. For Microsoft Azure, they are rolled up into something called Premium, whereas Amazon offers you Operational Security and Custom Deployment on their Amazon Web Services.

Another thing that differs between the different options is, of course, price. It's hard to do a good estimate on the total price, but Databricks offers a calculator that will at least give you a hunch of where you'll end up. You also get to see the factor between the different tiers. You can find the pricing information at https://databricks.com/product/pricing where you can look at the numbers for your cloud provider of choice.

At the time of this writing, it is $0.07 per Databricks Unit (or DBU for short) for Data Engineer Light, $0.15 per DBU for Data Engineer, and $0.40 per DBU for Data Analytics. These prices are the same on both platforms, but they only cover for the Databricks part. The cloud resources are extra, which is where the variability comes into play.

As an example, I calculated what it would cost to have four instances running 8 hours a day for 20 days in a month on AWS. I picked the m4.2xlarge machines with eight CPUs and 32 GB of memory. The result on AWS was $67 for Data Engineering Light, $144 for Data Engineering, and $384 for Data Analytics – a very reasonable price for the computing power you get.

Databricks on Amazon Web Services

Databricks was first introduced on the Amazon Web Services and therefore has the longest experience running there. It's also where the community edition runs. AWS is the biggest cloud provider and therefore a solid choice used by lots and lots of companies.

In theory, there are two ways you can get Databricks running on AWS. But while it is possible to buy Databricks of the AWS Marketplace, it's better to do it directly from Databricks. You'll get more options, and you get the free trial with just a button click. But, by all means, if you want to do it directly at AWS, it's possible. That's not the route we'll go here though.

Before we start, note that this process requires a credit card and that you'll automatically be enrolled into a paid tier of the service once your two free weeks are over. So if you just want to test it, make a note of the end date in your calendar so you don't end up overspending. It's very easy to quickly get very large bills if you configure huge clusters, process tons of data on them, and then on top of everything forget to shut them down or have them stop automatically.

You also need to have an AWS account. That is where Databricks will spin up the servers and install the software. If you don't have this already, head over to https://aws.amazon.com/ and create one. Note that Amazon frequently has offers for some free resources for new users. A free Databricks trial and free AWS computing power is the perfect combination to get started with large-scale data analytics.

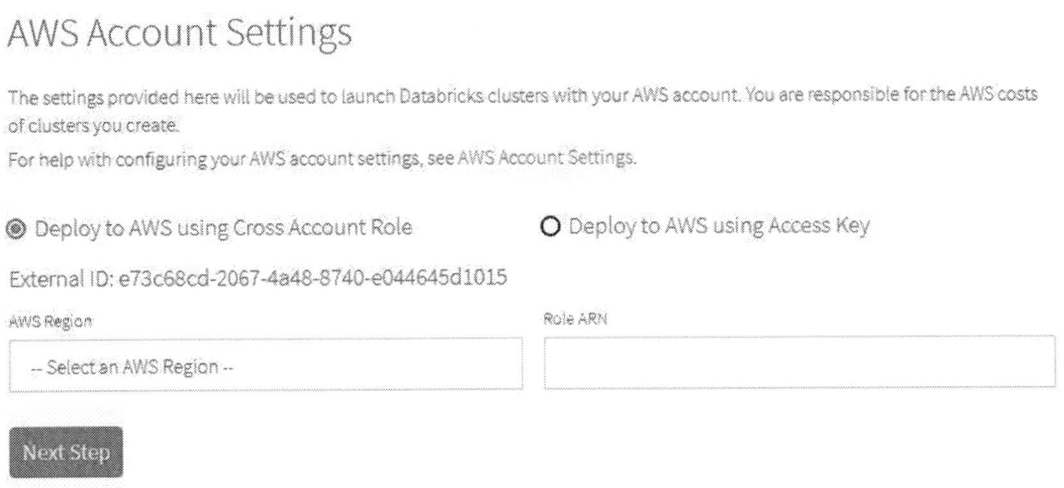

CHAPTER 3 GETTING STARTED WITH DATABRICKS

AWS Account Settings

The settings provided here will be used to launch Databricks clusters with your AWS account. You are responsible for the AWS costs of clusters you create.

For help with configuring your AWS account settings, see AWS Account Settings.

◉ Deploy to AWS using Cross Account Role   ○ Deploy to AWS using Access Key

External ID: e73c68cd-2067-4a48-8740-e044645d1015

AWS Region   Role ARN

[-- Select an AWS Region --]   [   ]

[Next Step]

Figure 3-4. *Using a cross-account role is the preferred and default option*

When you have the prerequisites done, go to www.databricks.com. Then continue by clicking the TRY DATABRICKS button on the Databricks main page. Click the GET STARTED button in the Databricks Platform – Free Trial column. Fill in the questionnaire and make sure you don't miss any of the required fields. Check your email for a confirmation link and click it. Set the password.

Next up, you need to type in the billing details. Same deal here. Type in all your information and continue on to the AWS Account Settings. This is where you'll connect the Databricks installation to Amazon's cloud service. You can choose to do this with a cross-account role or an access key. If you don't have strong reasons against it, use the first option (Figure 3-4).

Now you need to create the role in AWS. Copy the external ID string from the Databricks page and continue over to your AWS Account Console. Find the Identity and Access Management (IAM) view. Click Roles and then Create Role. Select the Another AWS account. In the Account ID box, type in 414 351 767 826 but without the spaces. I added them just to make it easier to read. This is the Databricks ID within the Amazon Web Services system.

Check the Require external ID box, and in the new box that appears, paste the external ID that you copied on the Databricks site earlier. Click Next until you get to the Review page. Here you pick a name for your new role, for instance, Apress-Databricks, and finalize the role creation process.

CHAPTER 3 GETTING STARTED WITH DATABRICKS

Next, choose your new role in the overview list. Follow that up by clicking Add inline policy. Select the second tab, named JSON. Here you need to insert a larger piece of text. As it's very long, I won't type it out here. You'll find it in the online download section for this book.

```
{
    "Version": "2012-10-17",
    "Statement": [
        {
            "Sid": "Stmt1403287045000",
            "Effect": "Allow",
            "Action": [
                "ec2:AssociateDhcpOptions",
                "ec2:AssociateIamInstanceProfile",
                "ec2:AssociateRouteTable",
                "ec2:AttachInternetGateway",
                "ec2:AttachVolume",
                "ec2:AuthorizeSecurityGroupEgress",
                "ec2:AuthorizeSecurityGroupIngress",
                "ec2:CancelSpotInstanceRequests",
                "ec2:CreateDhcpOptions",
                "ec2:CreateInternetGateway",
                "ec2:CreateKeyPair",
                "ec2:CreatePlacementGroup",
                "ec2:CreateRoute",
                "ec2:CreateSecurityGroup",
                "ec2:CreateSubnet",
                "ec2:CreateTags",
                "ec2:CreateVolume",
                "ec2:CreateVpc",
                "ec2:CreateVpcPeeringConnection",
                "ec2:DeleteInternetGateway",
                "ec2:DeleteKeyPair",
                "ec2:DeletePlacementGroup".
```

Figure 3-5. *The policy information is also available at docs.databricks.com*

Get it from the Web, paste it into the textbox, and then click Review policy (Figure 3-5). Give the policy a name and then click Create policy. Copy the role ARN in the summary screen that follows and head back to the Databricks site.

CHAPTER 3 GETTING STARTED WITH DATABRICKS

Pick a region and then paste the role ARN you just copied from AWS in the right-hand box. The next step will move you forward in the process. You have now connected your AWS account with your Databricks account and made sure all the necessary privileges are available. Unfortunately, you're still not done. You still need somewhere to store your data.

Go back to AWS. Find the S3 service and create a new bucket. Make sure it's in the same region as you mentioned when Databricks asked for that information. Then type that name in at the AWS Storage Settings page at Databricks (Figure 3-6).

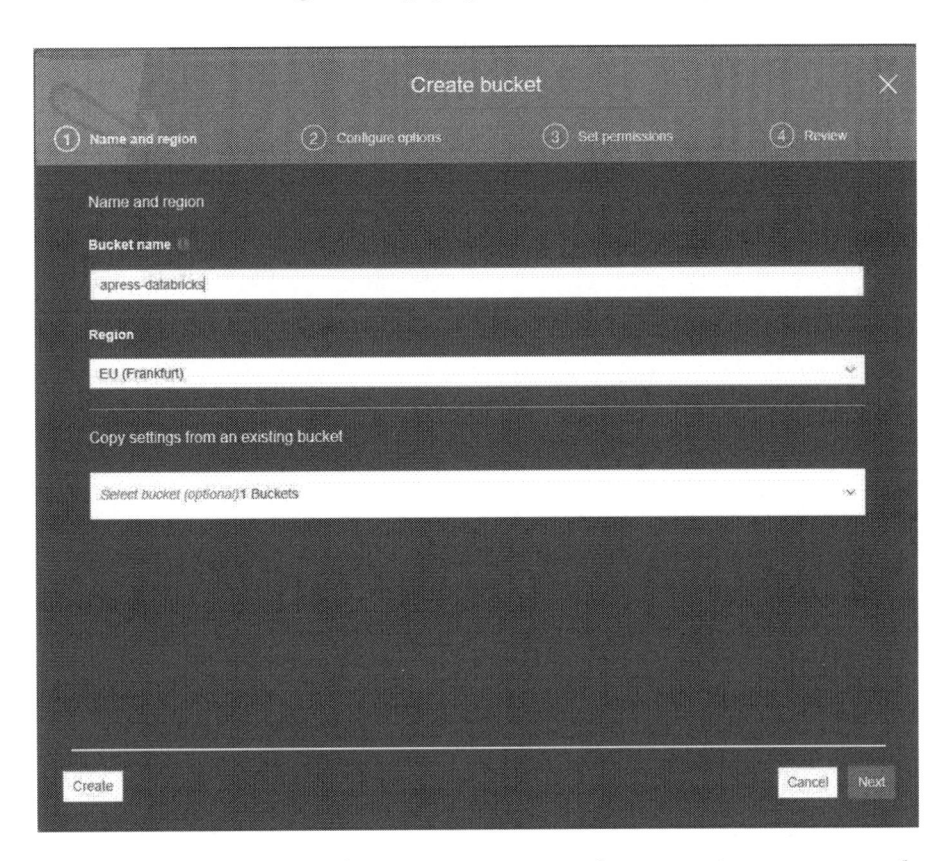

Figure 3-6. Always try to use descriptive names when creating pretty much anything in the cloud

Generate and copy the policy. Go back to your S3 bucket on AWS. Click the name and select the Permissions tab, followed by the Bucket Policy tab. Paste the policy you copied into the text field and save it. At the Databricks page, click Apply Change. Finally, accept the terms (after having read them, of course) and click Deploy.

35

After a rather long wait, up to 30 minutes, you'll have your Databricks installation done. Then you'll have a link on the Deploy Databricks page. Click it to get to your very first workspace.

Note: If you already have a trial account, you can go to the main page and click the Upgrade button up to the right. This will take you to the same overview page as mentioned earlier. The rest of the steps are identical.

Azure Databricks

If you think there are too many steps in the AWS setup, you'll be glad to know the Azure one is way shorter. Also, there's only one way to deploy it – through the Azure Portal. So you actually won't go through the Databricks website at all. Instead, head over to portal.azure.com.

Use your account to log in and create one if you haven't already. Just like with Amazon, there are frequently offers for some free resources available for newcomers. Either way, make sure you get to the main portal view.

Once you're there, search for Azure Databricks in the search box at the top. Select it once it comes up and click the Create Azure Databricks Service button in the middle of the screen.

In the setup screen, there are a few things you need to select. The workspace name is for yourself, the subscription needs to be connected to a credit card, and the location should be close to you physically.

You can create a new resource group or use an existing one. Cloud architecture strategies are outside the scope of this book, but if you're not sure, you should use a new one. The last option is for adding Databricks to one of your own virtual networks. This too is outside the scope of this book, but well worth looking into if you have cross-network communication needs.

CHAPTER 3 GETTING STARTED WITH DATABRICKS

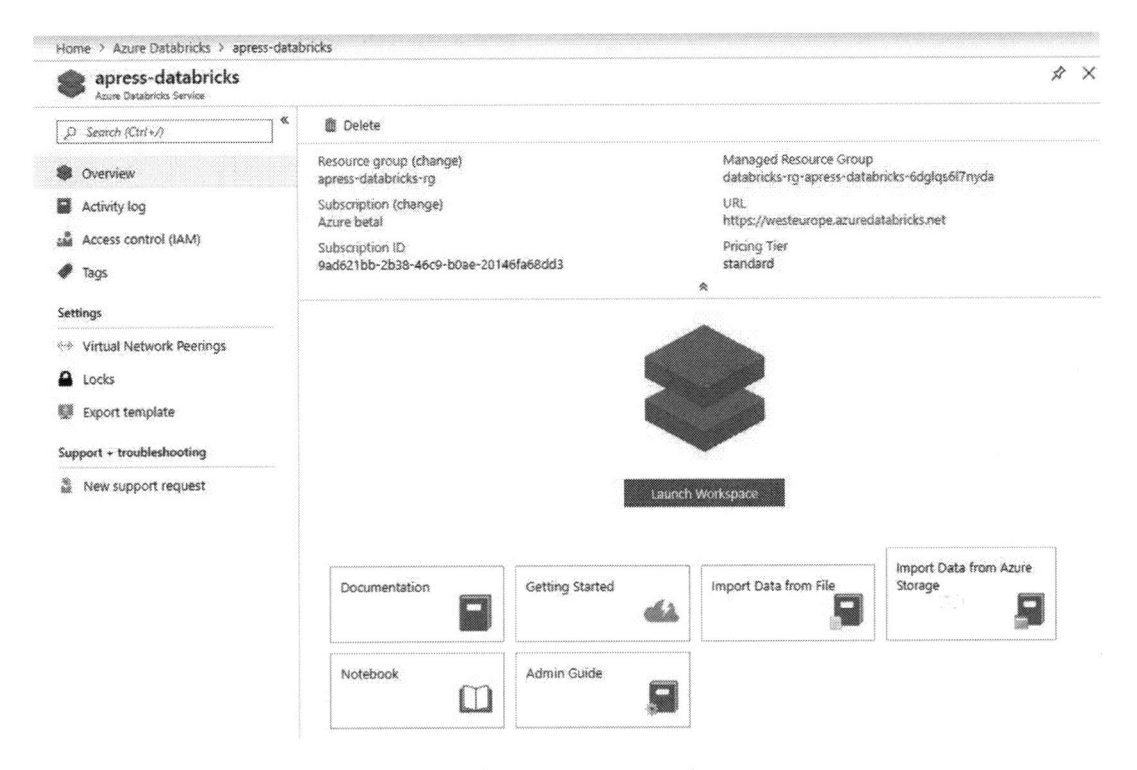

Figure 3-7. Configuring Databricks on Azure is a breeze

Once everything is filled in, just click Create. You'll see the progress bar in the Notifications tab. Wait a few minutes and then click Refresh. Then you'll have the new workspace in the list. Click it to go to the detail page (Figure 3-7). Click Launch Workspace to start Databricks.

That's it, quite a bit easier than doing the same thing with AWS. So if you're totally new to both the cloud and Databricks, I would recommend Azure for this reason. Getting up and running is just so much faster on Microsoft's solution.

It should be mentioned that Microsoft actually bought a small stake in Databricks in one of their investment rounds. This might be a reason for their apparent focus on this specific product even though they have other similar tools in their portfolio.

Summary

In this chapter, we went through the different versions of Databricks. We also got both the community edition and the commercial editions up and running on both of the available cloud platforms.

Now that we have our Databricks system ready to go in your preferred cloud, it's time to get familiar with the environment. Fortunately, this is exactly what we'll discuss in the next chapter.

CHAPTER 4

Workspaces, Clusters, and Notebooks

Finally, you have Databricks up and running on whatever your platform of choice is. For the most part, it doesn't really matter if you're on AWS or Azure from a user interface standpoint; Databricks looks the same.

When working with Databricks, you'll interface with workspaces, clusters, notebooks, and several other things. In this chapter, we'll go through them all, get familiar with them, and start testing them out. These are the fundamentals you need to find your way around the Databricks UI.

What you have created so far is what Databricks calls a workspace. It's a complete work environment that is isolated from other workspaces. So when you are setting up clusters, coding notebooks, and administering users, it's all limited to that one specific workspace.

Let's dig into what you can do in your environment. We'll start by just looking around and getting a feel for how the tool works. It's actually kind of amazing that so much complex technology can be presented in such an easy way.

Getting around in the UI

When you start Databricks, you are met by a general start screen (Figure 4-1). To the left, you have a small toolbar with a few buttons connected to features in the tool. On the top bar, you have links connected to your cloud account. Just beneath that and to the right, you have a quick link to the documentation and another one to your user account. Finally, you have a start page with a bunch of quick links to different functions in the middle.

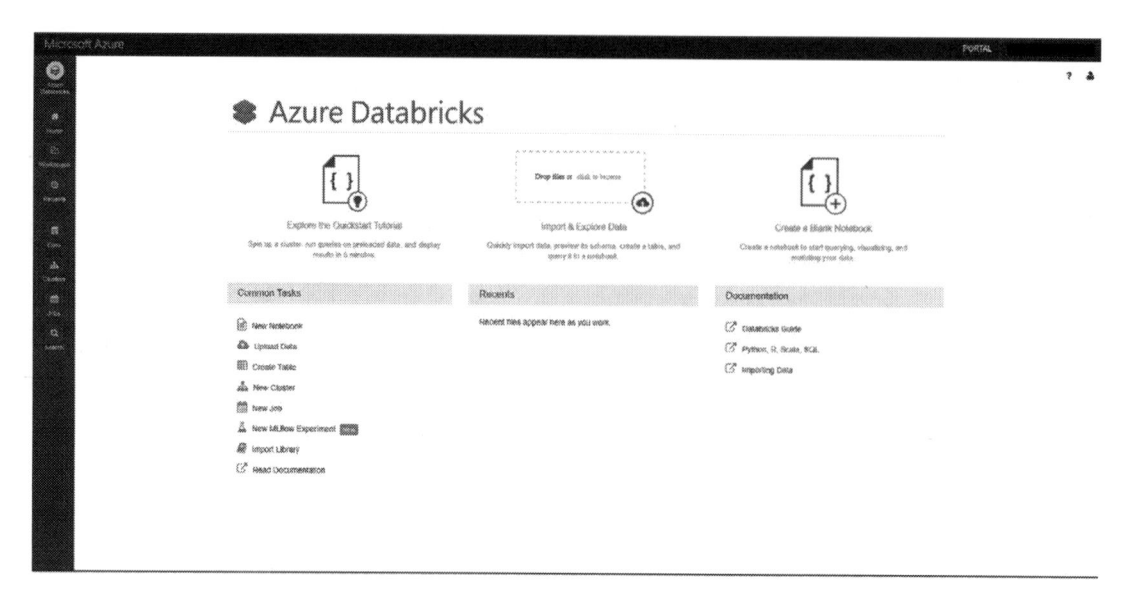

Figure 4-1. As Databricks is a cloud service, expect the UI to change from month to month. This image is from December -19.

Let's start with a few notes on the middle section, even though it's just a bunch of shortcuts. On top, you have links to the quickstart guide, the import feature, and notebook creation. We'll come back to that later. You also have common tasks in the first column, recent files in the second, and links to the documentation in the third – all pretty self-explanatory.

While you can do all the things available here at different places, I frequently come back to the start page to create a blank notebook. It's also a good place to pick up where you left off, considering that you can see your recent files here as well.

Mostly though you'll be using the left bar to actually access features and documents. It's available at all times, and most buttons will not transport you to another page, but rather just open a box on top of the existing page. Unfortunately, there's no indicator of which button opens a box and which replaces the workspace, but you'll quickly learn which one does what.

The top button, Databricks, will just take you to the start page where you already are. Both Home and Workspace will open the folder structure where you keep your files. The difference between them is that the Home button will take you to your personal root folder, while Workspace will return you to the place you were when you last closed the bar.

We'll talk more about this folder structure later, but there are a couple of interesting things in here worth mentioning already now. First, at the root you have folders containing documentation and training material. Second, at the same level, you have a Shared folder. While you don't need to use it for shared material, it's a good place for actually doing so as it's there by default.

Recents are, as you might imagine, a list of files and views you recently visited. Unfortunately, it's just a list of links. You don't get any metadata in this view, so you for instance don't know if you used the document last week of last year.

Next up is Data. This one you will probably use a lot. It lists all databases (or schemas, if you prefer that nomenclature) and tables in that database. You only see what you've stored in the metastore, so text files on your data lake won't appear here. We'll talk more about this in the next chapter.

The following two buttons are different in that they actually open new pages in the main view. So if you're in a notebook and click Clusters of Jobs, you'll replace what you were working with in the main work window. If you do it by mistake, the browser back button will help you come back to where you were.

Clusters is where you define and handle the underlying engines of Databricks. This is where you tell the system how many processing cores and memory you need. You can switch between clusters and pools at the top of the main view. We'll play around with this in a bit.

Jobs, which we will investigate in depth in Chapter 11, is where you schedule your finished products to run either again and again or at a given time. You define what needs to be done, which cluster should do it, and on what schedule. Then Databricks will make sure that your jobs are executed on time. You can also track what happened here.

Finally, Search will let you search in the folder structure for a document, folder, or user. It's pretty nifty if you use the same prefix for files connected to a project as you can then use subfolder and still get everything in one list.

Note that if you want to find stuff in the documentation or the Databricks forum, you can instead click the question mark up and to the right. The result will open up in a different tab in your web browser.

About navigation. It's easy to miss a few features here. Note that there are arrows halfway down for moving back and forward. There are also small handles along the edges you can grab hold of and expand the size of the windows – only in the middle though, next to the arrows. At the top right in the box, you can click the pin to lock down the box and keep it open, moving the work area to the right. None of these features are obvious. Databricks still has a little bit of work to do in the user interface department.

There's also another set of features hidden behind the icon with a picture of half a human, up to the right. This is your personal profile, and if you click it you see a few entry points you will use less frequently in your day-to-day data exploring.

User administration is where you can set up your version control handling, get API keys, and define a few personal preferences in how the notebooks should behave. The keys and the version control handling we'll come back to later in this book.

If you have administrative privileges, you'll also see the Admin Console. This is where you set up users, groups, and security options. It's also where you go to clear out logs and empty the trash can. The last part you'll hopefully not have to do frequently as it's mostly used to clear out information that shouldn't be in the system anymore. For instance, this can happen with a data protection request.

If you have multiple workspaces, you get a list of links at the bottom of the list. It's a quick way of moving from one project to another. Unfortunately, you don't jump back and forth like a window change on your computer, but rather you just get a new tab. Still, it's a good shortcut to have.

You can also log out. This is probably not something you'll do frequently, but if you are borrowing a computer, it's a good idea to make sure you disconnect from Databricks properly as to not let someone else access data using your profile.

Those are the basics of the user interface. There are of course a lot of details, and we'll visit many of them throughout the book. Hopefully you've seen enough to find your way around the system. So let's start using the different components. To be able to do anything, we need the engines to purr, so let's get a cluster up and running.

Clusters: Powering up the engines

The core of Databricks is of course the processing power you can spin up. It's required not only to run code but also to attach to the underlying Databricks File System. Building a cluster is as easy as typing in a name and clicking the Create Cluster button. There are a lot of options, but the default settings are actually good enough for most minor use cases and a good place to start.

In the following image (Figure 4-2), you see a page where only the Cluster Name field has been changed. Everything else is set to default, which means you'll get one of the smallest clusters you can build. It's still quite powerful though, as you'll see.

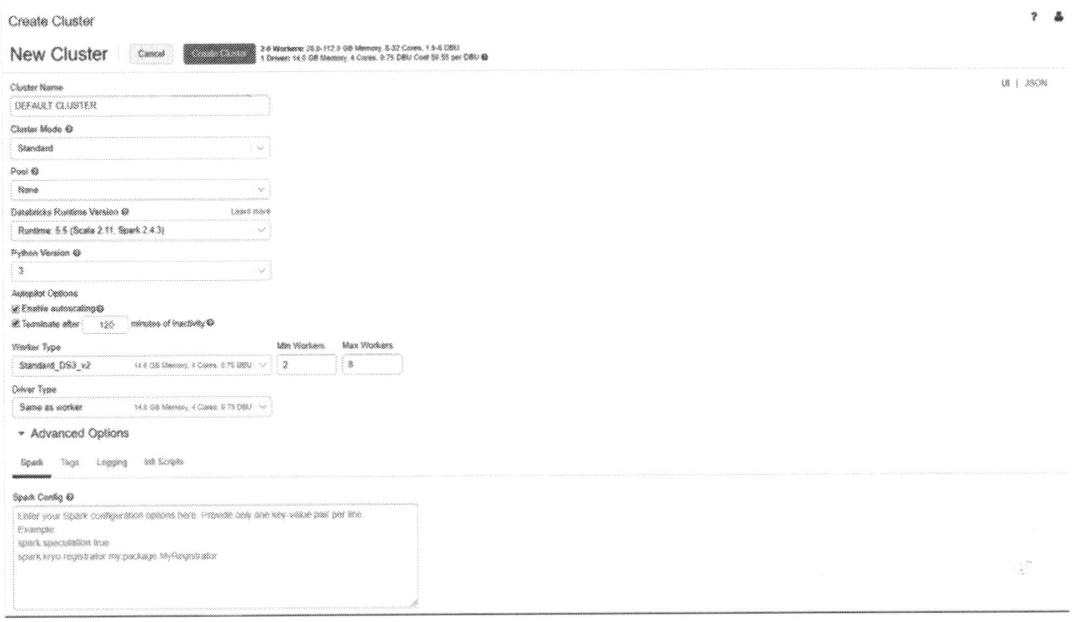

CHAPTER 4 WORKSPACES, CLUSTERS, AND NOTEBOOKS

Figure 4-2. *Note that options are slightly different between the different platforms. The image is from an Azure setup*

Let's look at the different options, from the top. In Cluster Mode, you decide if this is a Standard or High-Concurrency cluster. You want the High-Concurrency option if you want to have a large number of users using the cluster at the same time. This is not ideal, but is sometimes necessary. Most importantly, you'll need it for Table Access Control, as we'll discuss in Chapter 11.

Then you decide if you want to put the cluster in a pool. This is a fairly new concept for Databricks. The idea is that you can keep resources around in a pool to speed up starting clusters. This is good as starting clusters is otherwise slow, but keep track of costs as this option is securing machines in the cloud. Databricks, however, does not charge anything for the idle instances in a pool.

The runtime version is pretty straightforward. You pick the Databricks/Scala/Spark combination that works for you. Normally you should be able to run with the latest version, but if you want to be sure the code will work the same over time, it might be good to pick a long-term supported version, or LTS as it's abbreviated in the list. Be aware that older versions of course lack features newer ones have.

Python version, if the option is available, should be 3. Seriously, please do not keep using 2. There's no excuse anymore, and the old version is actively being phased out. If you're still in love with 2, get over it. But sure, you can pick 2 if you really have to for older

43

versions of the Databricks runtime. But don't. The community has decided to move over, and so should you. From version 6, it's not even available anymore.

The next four options are tightly connected. Driver Type defines the power you want to have on the driver node. If you expand the view, you'll see there are a ton of options with different settings for memory and processor usage. Some of them even offer graphical processor units, of GPUs, primarily for deep learning work. For every category, you can also click the "More" option to get a silly long list.

Worker Type is the same thing, but for the workers. The difference is that you can define the number of workers you want to have. There'll be just one box if you don't have Enable autoscaling checked and two if you have it checked, one box for the lowest number of workers and one for the highest number of workers. Databricks will automatically scale up and down depending on the current load if you ask it to. This is good from a cost perspective, but be aware there is a bit of a lag with scaling. Usually this is not a problem, but in some cases it's frustrating.

One of the really great features in Databricks is a simple little checkbox called Terminate after minutes of inactivity. If you spin up a big, non-autoscaling cluster and forget about it, you'll remember it when the bill comes. This option lets you have Databricks kill the cluster after a given number of hours of inactivity – a real cost saver.

There's a lot to change under advanced options as well. We won't cover all the details here, but we'll come back to them throughout the book. Just be aware that this is the place you'll find them.

Finally, note that there are a couple of strange UI and JSON links up to the right. You can actually set the cluster up and save the configuration file. Or edit the text version directly if you want to. This is useful when you want to copy identical settings between clusters.

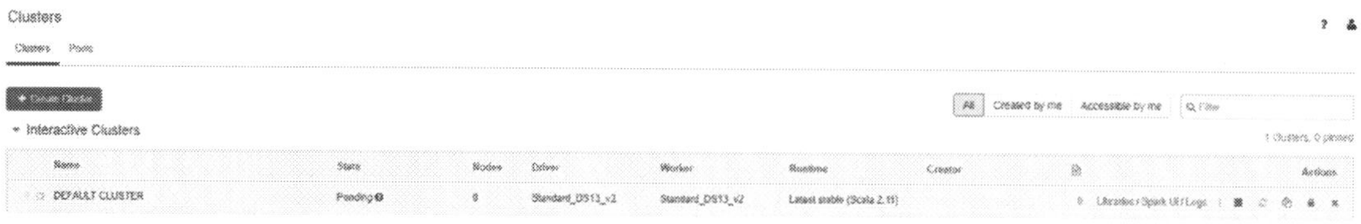

Figure 4-3. The green, spinning circle to the left shows the cluster is on its way up

Once you've configured a cluster, it'll show up in the overview list (Figure 4-3). To the left you see the name, the status, the types of machines, and the number of nodes. On the right-hand side, you have a few control options, including buttons for starting, stopping, and restarting the cluster. This is a place you'll visit frequently as you work with Databricks.

To continue, set up a default cluster and wait for it to start, which it will do automatically at creation. You'll need it to be active for the next section to work as you need a cluster to look at the data view. If you pause here, remember to start the cluster before you continue to the next section.

Data: Getting access to the fuel

This is your view into the data you have available in the metastore. The first column lists all databases you have available, and when you click one, you'll see all tables in the next column. If you click one of the tables, you'll get the details about that table in the main work view.

Until you've created your own databases, there will only be one database available, named "default." It's there, but you shouldn't really use it much. It makes much more sense to keep your data in a database with a given name. This is true even if only you will use the workspace. Keeping tables together will help you remember what they were for when you come back to them in the future.

In this pop-up view, you also have an Add Data button. If you click that one, you'll open an import view where you can upload data in a semi-graphical way. The web interface isn't great, but it's a direct way to get data into Databricks from several sources. It's especially useful for small, basic datasets such as comma-separated values (CSV) files. Let's try to pick something from your client.

Start by downloading meat_consumption_kg_meat_per_capita_per_country.csv. Then make sure you have Upload File selected. Drag the file onto the designated space or click Browse and find the file on your computer. The file will transfer into the cloud. Click Create Table with UI, select your cluster, and then click Preview Table. You'll now get a preview of the data in the view and a few options on the side (Figure 4-4).

CHAPTER 4   WORKSPACES, CLUSTERS, AND NOTEBOOKS

Specify Table Attributes							
Table Name	_c0	_c1	_c2	_c3	_c4	_c5
meat_consumption_kg_me	STRING	STRING	STRING	STRING	STRING	STRING
Create in Database	COUNTRY	BEEF	PIG	POULTRY	SHEEP	TOTAL
default	ARG	40.41400058	8.24187459	36.4689953	1.174247186	86.29911766
File Type	AUS	22.8010372	20.25072536	42.00750521	7.423454044	92.46272161
CSV	BGD	0.885267859	5.14E-04	1.223173534	1.163676301	3.272631248
Column Delimiter	BRA	24.15640871	11.20721696	39.36312514	0.393513724	75.12026453
First row is header	BRICS	4.269081407	15.79587836	10.29847417	1.654767905	32.03820184
Infer schema	CAN	17.37132968	15.74658647	34.15846671	0.81704465	66.09342751
Multi-line						
Create Table						
Create Table in Notebook						

Figure 4-4.   Once you get used to importing data, it's often easier to do directly from a notebook

We can see immediately that there's a header and that there are different types of columns, not just strings. Mark the First row is header and Infer schema checkboxes. Wait a few seconds, and you'll see that the headers are now correct as well as the data types. If you want to change anything manually, you of course can. Finalize the work by clicking the Create Table button.

Now if you go back to the Data sidebar and click the default database, you'll see the table to the right. Click it to see the details and a few sample rows. Now that we have the cluster running and the data in place, it's time to start looking at it. We do that in our notebooks.

Notebooks: Where the work happens

So the actual data processing work, including the data import, is done in notebooks – at least for the most part. As you'll see later, you can actually connect to Databricks from external tools and use it as an Apache Spark engine, but notebooks are the main way to talk to the system.

Currently Databricks supports four different languages in the notebooks: Python, R, Scala, and SQL. The first and last of those are the best supported ones in terms of security features, while Scala is native. Still, they're all good, so you can use the one you're most comfortable with unless you need the extras. We'll go through the differences in Chapter 10.

Let's try one out. Go to the start screen by clicking the Databricks logo up to the left. Then click Create a Blank Notebook. Name it Hello World and use SQL as the primary language. After a few seconds, the work area is created, and you'll be placed in the first textbox of the notebook (Figure 4-5).

Figure 4-5. Note that not all keyboard shortcuts work on international keyboards

If you look around, there's a bunch of information and features available. You have the name of the notebook, followed by the main language, at the top. Below it you see the connected cluster (and an option to change cluster), a few menus, and then another set of menus. The main part is the input window, or the cell as it's called. This is where you enter your code.

Type in the following command. Then click the play button on the right-hand side in the cell. This will open a dropdown list. Select Run Cell. This will execute all the code in the cell, in this case only a single command:

```
select * from meat_consumption_kg_meat_per_capita_per_country_csv;
```

The command is sent to the cluster where it's processed, and then the result comes back to you in a neat grid. You have a scroll bar to the right, and by grabbing the small arrow in the bottom-right corner, you can change the size of the cell.

While this is nice, clicking two buttons with a mouse every time you want to execute a command is a bit inconvenient. Let's learn a shortcut. Make sure you have the cell selected, hold down the Ctrl key, and then press the Enter key. This will execute the command. Much easier than to click the buttons, right?

Another little trick is to use auto-complete. If you type in the first few characters and then press Tab, Databricks can sometimes help you out. In the preceding example, just type in "meat" followed by pressing Tab, and you should automatically get the full name written out. Very helpful.

47

Having just one cell is a bit limiting though. Let's create another one. Move the mouse pointer to the lower middle of the cell. A plus sign will appear. Click it. Type in the following command and execute it:

```
select count(*) from meat_consumption_kg_meat_per_capita_per_country_csv;
```

That's the power of cells. Basically you can run commands one after the other with the output in between. So if you have one part done, you can leave that be while you work on the next part. You don't need to run the cells in order if you don't want to, but it'll be easier to understand what's going on if you do so.

You can of course remove cells. Just click the x button in the top-right corner of the cell. The downward-facing arrow gives you commands to, among other things, move cells around if you want to change order. There are a lot of small features you'll discover along the way in this book.

There's one really cool feature you should be aware of already now. This notebook has SQL as the primary language. If you want to run Python here, you still can. You just need to let the cell know you want to change language. You do that with what's called magic commands. Create a new cell and type in the following lines of code:

```
%python
df = spark.sql('select * from meat_consumption_kg_meat_per_capita_per_country_csv')
display(df)
```

The first row, %python, tells Databricks that this cell will execute Python code instead of SQL. It works just as well with %scala, %r, and %sql for the different languages. There's also %sh for shell script, %fs for file system, and %md for markdown, for documentation. You'll come across these again later, but let's try writing some information to remember what we've done. Try adding a cell at the very top. Then write this in the new cell:

```
%md
# Hello World
```

Now you have a headline at the top. Nice. It's always a good idea to fill your document with comments and thoughts. This is true even if you're the only one who'll ever be looking at it. Your future self won't have the same perspective you have right now. Help yourself out and document thoroughly.

Let's add a little graph as well. In the first cell we ran, locate the graph button down to the left in the cell. Click it. The result turns into a bar chart. If you prefer the grid view, just click the button to the left of the chart button. The downward point arrow on the other side of the Plot Options downloads the result in a CSV format. Good to have if you want to play around with the result in Excel or QlikView.

As you might imagine, this way of working is exceptionally good when working with data. You pick some information up, mess with it a bit, look at it, mess with it a bit more, look at it again, draw a few graphs, run an algorithm on it, and so on. Doing it one cell at a time makes it so easy to create a repeatable flow. When you're done, you can run everything from the top to the bottom. Add documentation between the cells, and other people can follow your reasoning as well.

There is one thing you need to keep in mind when working in notebooks, and that is state. When you run something, Databricks remembers the result. Store something into a variable, and you can pick it up in a later cell (or an earlier cell if you run them out of order). But if the cluster shuts down, it's all gone, and you have to build it up again.

If you want to clear the state manually, there's a button for that at the top. In the same dropdown list, there's also an option to just clear the result panes or to clear everything and rerun from the top. Run All will, not surprisingly, run all cells from top to bottom.

One more thing, if you click Revision history in the upper-right corner, you'll notice that all your work is being recorded. You can click any of the entries in the list and see the notebook from that period in time. Very helpful. Just click Exit once you're done looking at the old version, unless you want to keep it of course. If so, click Restore this revision.

Now you have the tools at your disposal to actually do some work. You just need some language to play around with the data. So let's see how classic SQL works in a Spark setting.

Summary

This chapter was dedicated to the basic interface of Databricks. You learned how to move around in the workspace and find all the core features available. Once we went through that, we created a cluster, opened a notebook, and ran some code.

We then looked at cells, the core input fields in Databricks. In that process, we learned about code execution, graphs, state, and much more. Also, we looked at different magic commands that help us document what we do.

Soon we'll learn more about the actual coding part. To do that, we need to have some data. In the next chapter, we'll look at how to get information into Databricks.

CHAPTER 5

Getting Data into Databricks

All the processing power in the world is of no use unless you have data to work with. In this chapter, we'll look at different techniques to get your data into Databricks. We'll also take a closer look at file types that you are likely to come across in your data work.

To get a better understanding of how data is stored in Databricks, we'll investigate their own file system, called Databricks File System or DBFS for short. With this knowledge, we'll look at how we can pull data from the Web, from files, and from data lakes.

Getting data is easier if you have continuous access to it. We'll look at how you can attach your cloud-based storage to Databricks and use it like a local file system. We'll show how this will make your life easier.

Finally, we'll also make a 180-degree turn and look at how to extract information out of Databricks. With this knowledge, you have everything you need to handle data in the raw format it's usually presented in sources in the wild.

Databricks File System

Handling data on an Apache Spark cluster is a bit special as the storage isn't persistent. The information needs to be saved onto a storage cluster so you don't lose everything at restart. Many local installations use Hadoop Distributed File System for this task. Databricks have their own solution.

At the center of Databricks data handling, they have their own file system. It's called, obviously enough, Databricks File System. It is distributed which gives you speed and resilience. It's also automatically created and connects to your workspace when you create a new one.

The Databricks File System is a persistent store. This means that everything on there will stay there even if you shut all the clusters down. So if you for instance want to export a file, it's better to do that into DBFS than into the cluster driver, which is cleaned up at shutdown.

Another benefit is that you can mount, or connect, other file systems from external sources. This makes it possible for you to attach a data lake folder and use the files in there as if they were stored on DBFS. That's very convenient for loading data, as you'll see.

While the underlying technology is advanced, you get an easy entry point to it. It pretty much resembles any other file system you've seen in the past. All the cluster technology is abstracted away. So files are stored in folders, and you navigate through them using simple commands.

You can access DBFS a number of different ways. You can use the Databricks CLI, a few different APIs, and Databricks' dbutils package. We'll mostly use the latter alternative in the next few chapters but will come back to the others later.

Navigating the file system

As it is the Linux operating system running underneath the surface, you can navigate the folder structure like on any other Linux system. If you are unfamiliar with the Linux file system, it works pretty much like most other file systems, with folders in which you put files.

From a very high level, the biggest navigational difference from Windows, which uses a drive as the root, is that Linux starts its structure at "/". That's the lowest level, the foundation on which everything else stands. Folders are then added on top.

It's normally a good idea to put stuff higher up in the tree if possible. Don't unnecessarily store things in the root folder. Although it will work just fine, you'll quickly clutter the file system and make it hard to find things. Nicely grouping stuff into folders with descriptive names is highly recommended. This is an example of a path in Linux:

/home/Robert/Databricks/MyProjects/Education/MyFile.py

To explore the Databricks File System, we also need to know how to navigate it. As mentioned before, there are many ways to do this. We'll use the easiest one to begin with – magical commands through notebooks.

If you have ever used Linux or any kind of Unix system, you'll recognize the commands we use here. It's not a coincidence. Databricks is running on Ubuntu, one of the most popular Linux distributions around. Most stuff you can do there you can do in Databricks as well. Now let's see what the file system looks like.

Create a notebook to get started. Pick Python as the main language. It won't really matter for the magical commands, but we'll look into the dbutils package as well, and Python makes it easier to use. Let's see what's available in the root folder:

```
%fs ls /
```

The percentage sign and fs tell Databricks that the command that follows should be run on the Databricks File System. And we start by running a list command, "ls". The first argument says which folder we want to investigate.

As explained in the preceding text, the slash indicates that you are looking at the root, or the bottom of the folder structure. What you'll find is a number of default folders: FileStore, which we'll come to in the next section, databricks-datasets, and databricks-results.

The databricks-datasets folder contains a large number of public datasets you can use to try features out. We'll use some of them later in this book, but feel free to investigate the folder to see what you can play around with without having to download external data.

When you work in your notebooks and decide to download the full result of a query, what's actually happening is that Databricks creates a comma-separated values file for you. That file is stored in databricks-results and is then sent to your browser.

Let's dig a little deeper into the file structure and see what else we can do. For instance, we can browse around in the dataset folder to see what is in there and maybe dig a little deeper into one folder:

```
%fs ls /databricks-datasets/
%fs ls /databricks-datasets/airlines/
%fs head /databricks-datasets/airlines/README.md
```

Here we looked first at the list of datasets and then listed the contents of the folder containing the airlines example. Finally, we used another command, head, to actually look at the README.md file in the folder. There are a bunch of other commands you can use as well, like cp, which is "copy" shortened:

```
%fs cp /databricks-datasets/airlines/README.md /
```

In this example, we copy the README.md file to the target folder /, or root. As you might remember from the preceding discussion, we shouldn't put files in the root folder. So let's use rm, which stands for "remove," to get rid of it. Be very careful with the rm command. It's possible to wipe out a lot of data and crash both the node and the cluster:

```
%fs rm /README.md
```

There are a few more commands you can use in this way. You can create folders, move files, and mount file systems for instance. If you ever forget about what you can do, just type out %fs by itself and run the cell to get a description.

If you do that, you'll also see that the magic commands are a shortcut. What's actually being executed underneath is the dbutils package. If you want to, you can use that directly. The way you do that is pretty similar to the way we've been doing it so far. You just have to use the package instead, with the files and folders being arguments:

```
dbutils.fs.ls("/databricks-datasets")
dbutils.fs.head("/databricks-datasets/airlines/README.md")
```

The benefit of using the dbutils package is that you can actually use it in code. So if you want to loop through all the files in a folder, you can do that based on the result set from an ls command, like this:

```
files = dbutils.fs.ls("/")
for f in files:
  print(f.name)
```

Alternatively, if you prefer the list comprehension version, you can write it in one line of code. Note that I'll not be using this construct in this book as it's slightly more confusing for newcomers to programming. Don't let me stop you from doing it if you prefer to, though:

```
x = [print(f.name) for f in dbutils.fs.ls("/")]
```

This code does the same thing as the preceding one and also returns the result to a variable, x. While it uses fewer lines of code, it is slightly harder to read even in this simple example. With more complicated loops, it quickly gets much worse.

The FileStore, a portal to your data

I mentioned earlier that the FileStore folder is special. Most importantly, everything you put here can be accessed through a web browser. So it can be used as a quick-and-dirty way to make data from your system available to the outside world.

You can actually do this from pretty much any folder in the file system, but it'll require an additional login. In this folder, stuff is pretty much just accessible. While this is a nifty little feature that we'll explore later in this chapter, it's important to remember that your data is exposed.

Another feature is that you can put objects here that you want to show in your notebooks. Maybe you want to add a company logo at the top of the notebook. You then put the file in a folder here, like /FileStore/images/logo.png, and access it using the displayHTML command. Note that FileStore turns into "files" in the path:

```
displayHTML("<img src = '/files/images/logo.png'>")
```

We'll come back to this feature later in the book. Right now, just remember that the FileStore is different than the other folders in the Databricks File System and that you can use it for moving data and storing assets.

Schemas, databases, and tables

While file is ultimately how data is stored in most databases, that's not how we normally interact with the data – at least not when it's structured. Instead, we work with tables within databases.

Let's clarify a few terms. A table in Databricks is the equivalent of an Apache Spark DataFrame. It's a two-dimensional structure similar to a traditional spreadsheet, with rows of data in shapes defined by the column.

The big difference compared to a spreadsheet is that the format of the data is strict. The structure of the data in the table is called the schema. It defines what the data looks like and what the columns contain, with limitations. It will, for instance, tell you that the third column is a numeric value with three significant digits. Data that doesn't adhere to these rules can't be added to the table.

Tables can be local or global. The main difference is that the global ones are persistent, stored into the Hive Metastore (which we'll talk about in a bit) and available across all clusters. Local tables, also called temporary tables, are only available in the local cluster.

> **UNSTRUCTURED DATA**
>
> Sometimes you'll hear talk about unstructured data. What people usually mean with this term is not that the data is truly unstructured, but rather that it can't easily be fit into a schema, at least not in a way that makes sense.
>
> An example is written text, like the one you're currently reading. There's structure to the language, but it's very hard to define it and impossible using schemas as used in traditional database operation.
>
> Databricks is very good at handling large amounts of text data, and for those use cases, you'll probably not use the tables described here but rather stay on the file system. If you do use tables, you'll probably just keep the raw texts in one column and use other columns for metadata.

The database concept is used slightly differently in different products on the market. In the case of Databricks, it's basically a logical container for collections of tables. You use it to separate tables between users, projects, or whatever you choose.

By keeping them isolated, you can use the same name for several tables. For instance, you might choose to create a database for project A and another database for project B. You can then create a table with the name of SALE in both of them. It'll be two different tables, belonging to two different databases.

Hive Metastore

All the information about the tables, columns, data types, and so on is stored in a relational database. This is how Databricks keeps track of your metadata. More specifically, Databricks uses Hive Metastore to store information about all global tables.

This is the database you use to see what columns are being used in a table and where the files are being stored. Note that it is only information about the tables that is stored in the metastore, not the actual data.

If you already use Hive Metastore in your organization, you can connect to that one instead. That's a topic outside the scope of this book though. Most users will probably not go down this path. If you want to, however, it's easily done. Take a look at https://docs.databricks.com/data/metastores/external-hive-metastore.html.

Apache Spark SQL is mostly compatible with Hive, and most functions and types are supported. There are some features that aren't supported, however. The work around this is continuously ongoing, and if you need to know the exact details, I recommend looking at the list at `https://docs.databricks.com/spark/latest/spark-sql/compatibility/hive.html#unsupported-hive-functionality`.

The many types of source files

Now that we know where Databricks stores the information, we can start to look at information that we'll normally see coming in. All the data that isn't created in Databricks has been created somewhere else and then transported and loaded. When doing the import, you need to know something about the file types.

While there are a huge number of options available, you'll normally just get in contact with a few common ones. In the data-driven world, you'll probably see Parquet, Avro, ORC, JSON, XML, and, of course, delimited files like CSV.

Delimited files are by far the most common way of moving data across systems even to this day. It's a classic solution to an old problem. You have probably come across CSV files, with the content being comma-separated values (hence the acronym.) It usually looks something like this:

```
Country,City,Year,Month,Day,Sale Amount
Denmark,Copenhagen,2020,01,15,22678.00
```

There are a few benefits with this type of data. Humans being able to read them is one. Clarity is an extension of that. Ease of adding data is another pro. Most importantly, pretty much any data tool in the world can read them, from Excel and Access to Oracle RDBMS and Qlik Sense, enterprise or personal. Just point to the file and load it.

Almost all tools also offer a way to write CSV files. For that reason, it's kind of a default choice. You ask someone for data, and what they can usually provide you with immediately is a delimited file. Just agree on the delimiter and away you go.

Unfortunately there are also many problems with CSV, or any delimited files for that matter. Compression needs to be handled manually, the data can contain the delimiter, you can't validate the consistency easily, and there's no schema built in – just to name a few. Loading these files is a pain. Avoid if possible. Although, you probably can't completely. Most data you'll find on the Web will be offered in a delimited format.

A slightly better alternative is to use JavaScript Object Notation, JSON. Like delimited files, they're easy to read, but they also add a little bit of description of the data. Schema support is in the format, but rarely used.

JSON is probably the most commonly used format for REST-based web services (and the oh-so-hot microservices). So if you connect to a data source on the Web, it's very likely that you'll get JSON objects back. It'll probably look something like this:

```
{
 "pets":[
  { "animal":"cat", "name":"Tigger" }
 ]
}
```

The problem is that JSON is frequently nested and not two-dimensional, making it slightly harder to parse if you want to save the data into traditional tables. Databricks makes it possible to create JSON-based tables, but using them is still slightly more cumbersome than basic DataFrames.

What goes for JSON mostly also goes for XML, eXtended Markup Language. It is, however, much harder to read by a human – not impossible, but definitely harder. The reason is that there are tags surrounding all data, making it verbose. I personally much prefer JSON:

```
<pets>
  <pet>
    <animal>dog</animal> <name>Bailey</name>
  </pet>
</pets>
```

Going binary

While JSON and XML might be better than CSV, they're not the best of alternatives for data transportation. You ideally want something that is space efficient and keeps the schema for efficient data load.

If you have a chance of choosing format yourself, there are three common ones you'll want to pick from in the world of Apache Spark. They are Avro, Optimized Row Columnar (ORC), and Parquet. All three are binary, as in machine-readable only unlike

JSON and CSV. They also contain the schema information, so you don't have to keep track of that separately.

Avro is different from the other two in that it is row based, just like most traditional relational database systems. While this might be a very good format for transporting data or streaming, you probably want to use something else when storing data for analytics consumption.

While the difference between row-based and column-based storage might sound esoteric, it is not. In many cases, the wrong option can lead to slower performance. The reason can be found in the underlying storage layer.

Data is being stored in chunks, or blocks. Assume that each block can contain ten data items and that you have a table with ten columns and ten rows. In a row-based solution, you'd have one row per block, whereas in the column-based solution, you'd have one column per block.

If your solution asks for all data for row 4, this would mean you only need to read one block in the row-based storage. You get block number 4, and all the information you need is in there. You've read a total of one block. Very efficient.

Let's say instead that you want to sum all the values in column 3. As the column data is spread out across all blocks, you need to look in all ten of them to get the total. Worst of all, 90 percent of the data you process is unnecessary as you only need one column per block.

For column-based solutions, the situation is reversed. As the column data is being kept together, you only need to read the column you want to sum up as to get the total, meaning one single block. On the other hand, you'd have to read all ten blocks to show a complete row.

As you can see, each solution can do one type of operation very efficiently and the other type less so. In reality there are a lot of smart tricks as to not make this as horrible as in the example, but overall the principle holds true. How you store your data matters.

Most analytics are done on a few columns at a time. This is the reason you won't see row-based storage very frequently in systems solely focused on analytics, like Databricks. Keeping rows together is much more common in operational systems where you frequently want to see or change multiple columns.

Both ORC and Parquet are column based and somewhat close to each other in technical terms, making the choice between them a bit harder. They both compress data really well, which is frequently the case for columnar formats (as data is frequently similar within the column). They are both fast.

My own testing on large datasets has shown Parquet, the default format, to be faster than ORC in Databricks. For that reason, I recommend using Parquet whenever you can. That said, it's never a bad idea to test the alternatives, especially as ORC has features that Parquet lacks. Creating a table with a different storage format is very easy to do, as you'll see in later chapters.

Alternative transportation

While Parquet files are the ones you want, if you can choose freely, it's frequently better to transport data in a different way. At least if you want to do this regularly. There are a number of tools that can transport data between many different source systems to Databricks. Just in the Apache ecosystem, you have Apache Sqoop, Apache NiFi, and Apache Kafka.

The benefits with these systems are many. For instance, it's easy to create flows, they support a large number of sources out of the box, and they offer solid tools for monitoring. It's a complete kit for extracting, transforming, transporting, and loading. All of this can be done in Databricks directly, but specialized tools do it better in my opinion.

That said, there are arguments for avoiding these tools. If you only have a few sources, run a lot of one-off data loads and feel more confident in Databricks than in the tools mentioned; you can keep using Databricks to load data, especially if you are willing to do a little bit of extra work. We'll get back to integration tools a few times later in the book.

Importing from your computer

Time to get working. For our examples in this chapter, we'll use data from the website Ergast. They keep track of everything that happens – and has happened – in the Formula 1 Circus, a little bit of data to play around with. Before we can look at the information, we need to import it.

So let's start by reading in some data. More specifically, let's import the results for the constructors. If you're not familiar with racing, constructors are how you refer to different teams, like Ferrari and McLaren.

Download the Formula 1 data from the Github repository and unzip f1db_csv.zip. There are a bunch of different files in the folder. This dataset contains a number of files that contain information about all Formula 1 races from the 1950s: drivers, constructors, results, and more. This is a very nice little dataset that is being continuously updated by Ergast.com. While we have a copy in our Github repository, you can get fresh data directly from Ergast. We won't be using all of them, but feel free to explore them if you're interested in motorsport.

Open one of the files – the smallest is usually a good choice – in a text editor. This will give you a sense of how the information you're looking at is structured. You'll notice two things immediately. The separator used is the comma sign and there is no header. This we need to remember.

Now make sure you have a cluster running. Then click the data icon to the left. In the upper-right corner of the panel, you'll see an Add Data button. Click it, and you'll get a new page dedicated to uploading information. Make sure you have the Upload File tab active.

Drag the constructor_results.csv over into the light-gray File box. Wait until it uploads the file and then click the Create Table with UI button that appears. We'll come back to the notebook alternative later.

Pick your cluster in the dropdown list and then click Preview Table. This will bring up a preview of the table and a number of options on the left-hand side. Databricks will figure out what we just saw ourselves and neatly place information in boxes with predefined, poorly named headers.

So let's go through the options. Change the table name to constructor_results. This is the name you will refer to later when you want to use the table. Choose to create the table in the default database. This is normally not a good idea, but we'll keep it easy here.

The file type is correct, so leave it to CSV. Column delimiter is also prefilled and correct. As you can see on the data, there is no header row, so we keep the First row is header option unchecked.

Infer schema we want though. Select it and Databricks will start working on the rows. This option makes the engine go through the file and make an educated guess about the contents. Instead of just having strings for everything, this gives us integers and doubles as well.

The guesses made for this table look good, but note that is not always the case. Also, this option can take a long time if there's a lot of data in the table. Specifying the schema is usually much faster. You can of course also choose to import everything as strings and fix it later.

Before we create the table, we also need to change the column names. Using the default _c0 isn't very good. Name the five columns, from left to right: constructor_result_id, race_id, constructor_id, points, and status. Then click the Create Table button.

This will create the table and show you a summary page with columns and some example rows. If you click the Data button and have the default database selected, you should see your new table to the right. Congratulations, first data uploaded!

Getting data from the Web

While uploading data worked well, it was a roundabout way of doing this. We went through our client for no reason. The source was on the Web, so picking it up directly from the source to Databricks would have been easier. Let's see a way to do exactly that.

The easiest way to get the data is by collecting it to the driver node in your cluster, doing whatever needs to be done, and then copying the results to the Databricks File System. So let's try that approach with the same file set as earlier.

You'll notice that we use the %sh magical command. That is for running shell commands. Everything you do will happen on the driver node which, as mentioned before, is a Linux machine. So you have a lot of options.

Like always, make sure you have a cluster up and running. Then create a new notebook with any main language. It doesn't matter what you pick simply because we will use shell scripts. You can also use whatever name you want, but ReadDataFromWebShell is as good an option as anything.

Working with the shell

There are a number of ways of getting the file we want. The easiest is to simply use the wget command. It is simple and fast and doesn't require any extras as it's already installed and ready to go on the driver node. In the first cell, type in the following commands:

```
%sh
cd /tmp
wget http://ergast.com/downloads/f1db_csv.zip
```

CHAPTER 5 GETTING DATA INTO DATABRICKS

The first row tells Databricks that we're running shell commands in the cell. Next, we move to the /tmp folder in the structure using the cd command. It's a good place to put a temporary file, but it doesn't really matter where you put it. As you might remember, nothing on the driver stays there across restarts.

Finally, we use the Linux command wget to pull the zip file from the Web and store it in the local /tmp folder. If you want to verify that the file was actually copied, feel free to list the content of the folder using ls. This should show you that the file is there:

```
%sh ls /tmp
```

Note that we typed out the command on the same line as the magical command. It's not necessary, it's just possible and a bit convenient if you only have one line of code. You should also be aware that your path will reset between running commands.

Next up, we need to unzip it. As we only want the constructors' data for this test, there's no meaning in expanding all of it. To get that file, we need to find out what the name is, so we have to list the contents first. Luckily we can both list and pull out a file using the unzip command with a few parameters. Let's try it:

```
%sh
unzip -Z1 /tmp/f1db_csv.zip
```

This will list all of the contents (you can use the -l argument instead if you want sizes as well). As you can see, the name of the constructors file is, conveniently enough, constructors.csv. So that's the file we want to pull out. Unzip comes to the rescue again:

```
%sh
unzip -j /tmp/f1db_csv.zip constructors.csv -d /tmp
```

Like before, we start with the magic command to indicate we're running a shell command. Then we run unzip to extract one file and store it in the tmp folder. Let's verify that it all went well. The Linux list command, ls, can help us out:

```
%sh ls /tmp/*.csv
```

You should now see a list of all CSV files in the tmp folder. The asterisk replaces any amount of characters so the command will match any file ending with CSV. Most likely you'll only see the one we just extracted. If you don't see any files, you need to redo the preceding steps.

Next up, we need to move it from the driver node to the shared file system. For that we need the move command, mv. The Databricks File System is also connected (mounted, in Linux terms) to the driver automatically under the /dbfs folder:

```
%sh
mv /tmp/constructors.csv /dbfs/tmp
```

With this command, we're telling Databricks to move the file we just extracted from the driver to the Databricks File System (DBFS.) This makes the file permanently stored on the shared storage and available even after a cluster restart. Also, it'll be available in the UI for importing data.

Note that you could have downloaded the file directly to DBFS, but I wouldn't do it unless it's a big chunk of data. The risk is you'll fill the file system with junk. This doesn't have to happen, but it tends to. Doing the raw work on the driver is better for that reason.

To continue, click the Data button and then Add Data. Click the DBFS button and navigate to the tmp folder. There you'll hopefully find our file. Choose it and continue with the Create Table with UI button. The rest works as with the manual solution we did earlier. Name the columns constructor_id, constructor_ref, name, nationality, and url, respectively. Don't forget to define the proper schema either manually or by using the Infer schema option.

Basic importing with Python

Now we've tried reading data using both a fully manual process and a shell-based solution. But you can actually do all this in Python as well. Let's see what the same steps would look like using code. So create a new notebook. This time you need to pick Python. Name it ReadDataFromWebPython.

First of all, we need to get the file from the Web. This can be done in a number of ways. The easiest and Databricks-recommended one is to use urlretrieve. While that works just fine, I actually recommend against it simply because it's flagged as legacy in Python. So it could just be removed one of these days, and you'd have to go back and change your code. Instead, you can use the requests package:

```
from requests import get
with open(' /tmp/f1.zip', "wb") as file:
    response = get('http://ergast.com/downloads/f1db_csv.zip')
    file.write(response.content)
```

The requests module is preinstalled in the driver, so we don't need to load it. So the first thing we do is to import the get function from the requests module. Then we prepare a new file, called f1.zip. Then we pull the file from the Web and, in the next line of code, write it to the file system – in this case to the driver. If you want to do the work directly on the DBFS, /dbfs/tmp/f1.zip will work.

Next up, we need to look at the content, find what we want, and unzip the file. As we have the constructor standings and the constructors, it might be good to also add the seasons. This will give a bit of data to look at:

```
from zipfile import ZipFile
with ZipFile('/dbfs/tmp/f1.zip', 'r') as zip:
  files = zip.namelist()
  for file in files:
    print(file)
```

We start by importing the ZipFile function in the zipfile module. Once that's done, we open the file we downloaded and pull out the list of names. While we could just print the output of the namelist() function, we print it out in a nicer way, row by row. The file we want is seasons.csv. Let's extract it:

```
from zipfile import ZipFile
with ZipFile('/tmp/f1.zip', 'r') as zip:
  zip.extract('seasons.csv','/tmp')
```

The two first rows are identical to what we just did earlier. Then we just specify what to extract and to which folder. Hopefully this works fine for you. You can use the preceding Linux commands to verify – of course use Python for this as well:

```
import os
os.listdir("/tmp ")
```

As you probably can imagine, the os module contains a lot of functions for communicating with the operating system. One such function is to list the content of a folder, which is what we do on the second line. Hopefully you find your file in the list.

Next up, we need to move the file from the driver to DBFS. This can be done in multiple ways, but let's get familiar with one package you'll use a lot: Databricks' own dbutils. With this, you can do a lot of things on a file system level:

```
dbutils.fs.mv("file:/tmp/races.csv", "dbfs:/tmp/races.csv")
```

Now that you have the data on DBFS, you can use the same user interface as you did in the first example. Go to Add Data, but use DBFS as source. Go to the /tmp folder and pick your file. The rest is identical to what you did earlier.

But as we've now gotten into running Python, let's also try to import the data using code. This is much more efficient if you need to read a large number of files or new versions of your data over and over:

```
df = spark \
.read \
.format("csv") \
.option("inferSchema","true") \
.option("header","false") \
.load("dbfs:/tmp/seasons.csv") \
.selectExpr("_c0 as year", "_c1 as url")

df.write.saveAsTable('seasons')
```

This might be a lot to take in at once, but it's actually not that complicated. We use the built-in spark.read function. We define that we're reading a CSV file, infer the schema, and specify that there's no header in the file. Then we specify the file to load. The selectExpr part changes name from the default _c01 .. _cnn to our choices. We then write the dataframe to a table in a separate command.

If the code looks a bit confusing, don't worry. We'll soon get to Chapter 7 where we'll go through code like this in a more detailed manner. You'll soon get the hang of it if you don't already know it.

Getting data with SQL

With manual intervention, shell scripts, and Python, you can do all the steps. If you however have the CSV file ready, you can actually use plain old SQL to access the data directly (Figure 5-1). Use one of the preceding methods to extract races:

```
%sql
create temporary table test (year INT, url STRING) using csv options (path "dbfs:/tmp/seasons.csv", header "false", mode "FAILFAST");
select * from test;
```

With this command, we create a temporary table (so only available in this cluster) connecting to the CSV directly. The first two options are pretty self-explanatory; the path is where the file is located, and the header just informs Spark SQL not to use the first data row as column names.

For the last option, mode, you need to decide what you want Databricks to do with malformed rows. Permissive will try to let everything in, using nulls to fill up missing values. Dropmalformed just throws bad rows away. What we went with here, failfast, aborts immediately when there's bad data in the source file.

While working with CSV files in this way is pretty error prone, it's a pretty nifty way to quickly get some results without having to go through any data load process. If you plan on using the data a lot, you're better off creating a permanent Parquet-based table instead.

Mounting a file system

One great thing about the DBFS is that you can mount external file systems to it and easily access data as if it was locally stored. Mounting basically means you create a folder that links to a folder on another system – kind of like file sharing in Windows networks, only even more transparent.

Mount points are available to all clusters in the workspace and are even persistent across restarts. The best thing about them however is that you can store your actual data in a central place, like a data lake, and just attach them when you need them. This is an excellent way of sharing data between workspaces and even different software. Beware of the user privileges though as data in the share can be accessed by all users on a workspace if not set up correctly.

Let's look at how to mount Amazon S3 buckets and Microsoft Blob Storage to DBFS. These are the two most commonly used types of storage and probably what you will see most frequently.

Mounting example Amazon S3

Let's get started with Amazon S3. I should mention at the very top that the way we're doing it here is not the recommended way of mounting file systems in AWS. You should be doing this with IAM roles instead.

The reason I'm not showing that alternative is space. Setting up IAM roles isn't very hard, but there are many steps, and describing that here would be too much. Please check the appendix for the steps. What we show here is a quick-and-dirty solution that will work for your private projects.

Start by logging into the AWS portal. Click in the search box and type in S3. Pick the top option in the list, and you'll end up in the S3 buckets list. Click the Create bucket button and type in the bucket name you want to use. Instead of going through all the steps, just click Create on the lower-left side. You now have a file area.

Next up is getting the keys to AWS so you can connect from Databricks. Click your name, up to the right, and then My Security Credentials. This should take you to a page with multiple security options hidden beneath tabs.

Click the Access keys tab to expand it and then click the blue Create New Access Key button. When you get the pop-up window, click the Show Access Key link to see the keys you need. Copy both of them, or download them in a file if you so wish.

You now have everything you need to get going in Databricks. Head over to your workspace and create a new Python notebook. Type in the following code, but replace the access key, the secret key, and the bucket name with the values you used:

```
dbutils.fs.mount("s3a://<access key>:<secret key>@<bucket name>", "/mnt/your_container")
```

Execute this code in the cell, and unless you get any errors, you now have a view into your S3 bucket from Databricks. Everything you put there will be accessible from your workspace. You can even put your data files in AWS and bypass DBFS. Right now, however, there is nothing there. You can confirm this by running

```
%fs ls /mnt/your_container
```

You'll get a message saying no files can be found. To change that, you can upload data in the AWS user interface, but let's create a file there ourselves instead. Let's read one of the tables we've already created and save it to S3 using Python:

```
df = spark.read.sql('select * from seasons')
df.write.mode('overwrite').parquet('/mnt/your_container')
```

The first line pulls the data from the "seasons" table into a dataframe. Then we write that dataframe into our mounted folder in the Parquet format. Now if we run the ls command again, we'll see the file there.

CHAPTER 5 GETTING DATA INTO DATABRICKS

Remember, this solution is only ok to use for personal, small-time stuff. We take a few shortcuts here to show how it's done, but once you start using this for real, please use IAM roles or at least secrets. The latter we'll come back to in a later chapter. IAM roles are outside the scope of this book, unfortunately.

Mounting example Microsoft Blog Storage

Microsoft offers several types of storage in their Azure cloud. While data lake might be the most fully featured one, their blob storage offers a cheap way to store a lot of information. Like with Amazon S3, we'll use the mount command in Python to get this done.

There are many different ways to access the data. One of the easier ways is to use a shared access signature. This is easy to do and makes it possible for you to limit the time the data should be accessible.

Go to the Azure Portal and to the Storage Account view. Add a new storage and type in all the information you are asked for. When you're done, just click Review + create. There's no need to step through the different pages. Note that the name you choose will be referred to as the storage account name.

Once it's built, go to the main page of your storage account and click the Containers link, followed by the plus sign at the top. This will create a new container, which you can think of as a folder. Name it and continue.

Next up, click the Shared access signature link on the left-hand side. You'll now get a view where you can define what the temporary access should allow and for how long. The default is to allow everything for eight hours. This will work for our example, so just click the Generate SAS and connection string button. Copy the SAS token, which is the second one of the boxes that show up.

Now go back to Databricks and create a Python notebook. Name it whatever you like and then add the following code to the first cell. Be sure to replace the <container name> with the name of your folder, <storage account> with the name of the storage account, and <SAS key> with the string you copied earlier:

```
dbutils.fs.mount(
  source = "wasbs://<container name>@<storage account >.blob.core.windows.net",
  mount_point = "/mnt/your_container",
  extra_configs = { "fs.azure.sas.<container name>.<storage account>.blob.
  core.windows.net" : "<SAS key>"}
)
```

This will mount your newly created container to Databricks, under the folder /mnt/your_container. Just to verify that everything worked well, you can run a list command on it. There will be nothing there, so you only get an OK back if all is...well, ok:

```
%fs ls /mnt/your_container
```

As we created the shared access signature with full privileges, feel free to create an object in it from Python. You can, for instance, read one of the tables we created earlier and save it to the folder in a different format:

```
df = spark.read.sql('select * from seasons')
df.write.mode('overwrite').parquet('/mnt/your_container')
```

Now if you run the list command again, you'll find a Parquet file in the folder. If you go back to the portal and into the container, you'll also see the same content. So this is a good way to shuffle data back and forth, especially as you can actually just connect Databricks to the stored files and avoid moving it around. More on that later.

Note that it's better to use proper authentication through your Active Directory and adding privileges on the actual files and folders in your data lake. That requires quite a bit of setup that is outside the scope of this book. Look in the extra material on our site to find more information.

Getting rid of the mounts

While access to external data is great when you need it, too many mounts can easily clutter your file system. So if you don't need a file system any more, you should disconnect it. This is done using the dbutils.fs.unmount command:

```
dbutils.fs.unmount('/mnt/your_container')
```

If you mount file systems in a notebook every time you run it and in the end unmount them, you should also wrap an unmount command at the top of the notebook into a try/except clause. That way you'll avoid getting error messages. We'll talk more about this in Chapter 7.

CHAPTER 5 GETTING DATA INTO DATABRICKS

How to get data out of Databricks

Getting data into Databricks is all good and well. Sometimes you want to get it out of there as well. There are quite a few ways of doing this. What will suit your need depends on where you want the data and how large your dataset is.

Easiest is to simply use the notebook user interface. As we've seen in the last chapter, you'll get a few buttons on the bottom of the result cell. The one farthest to the left is a picture of an arrow pointing down. That's what you want here. Run the query to get your result and click the button to save it locally in a comma-separated format.

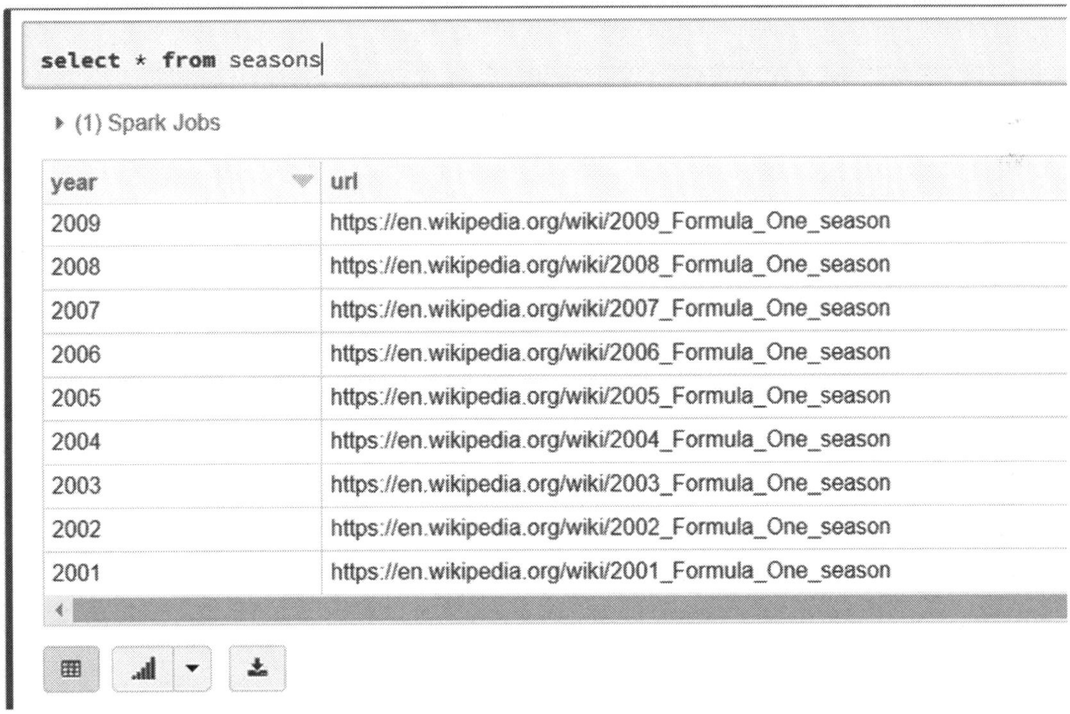

Figure 5-1. Reading everything from a table is often a good way to get a general sense of the format and contents

This method only works if you have a small amount of data. If it grows bigger, this method won't work. Instead you have to look at something else. Let's try a somewhat odd option that Databricks offers – exposing files via the Web.

As explained earlier, all the files you save in the FileStore folder will be available for direct download on the Web. This is very convenient and can be used for downloading really large files. It can also be used for small files, of course. Remember to be careful

CHAPTER 5   GETTING DATA INTO DATABRICKS

about what you put here as others can also download the file. People can be very clever about guessing filenames. Note that this is not available on the community edition of Databricks:

```
df = spark.sql('select * from seasons')
df.write.json('/FileStore/outdata/seasons.json')
%fs ls /FileStore/outdata/seasons.json/
https://westeurope.azuredatabricks.net/files/outdata/seasons.json/part-
00000-<some long string here>-c000.json
```

Note that you actually can access files in other parts of the file system as well by typing out the full path. This will however require you to log in to get the file, which is a good extra security layer. You don't want anyone having access to all your files.

Another, similar way of extracting data is to save it back to the cloud file system. This is frequently the best way of moving a large amount of data around. A good example is if you want to share data between workspaces. Best of all, we can use the same mounts as before. We've actually already seen how we can do that, but here it is again. Just make sure the file systems are mounted:

```
df = spark.read.sql('select * from seasons')
df.write.mode('overwrite').parquet('/mnt/your_container')
```

So far we've only worked with files, which is frequently the easiest choice. If you want to access the information from another tool, like Power BI or Qlik, it's a clumsy way of doing it. Instead of writing a file and then reading it, you can just access the data directly. You do this with either Open Database Connectivity (ODBC), Java Database Connectivity (JDBC), or a Spark driver. These we won't go through now. They'll show up later in the book.

Summary

In this chapter, we discussed the Databricks File System, looking at how data is stored and what's beneath the surface. We also navigated around the file system to get familiar with some of the more common commands.

We also introduced several ways of importing data into Databricks. Using different methods, we pulled data from our client computer and from the Web to our workspace and saved the result into managed tables.

CHAPTER 5 GETTING DATA INTO DATABRICKS

By mounting file systems, we saw how we could have access to data from both AWS S3 and Azure Blob Storage as if the files were in Databricks. This proved to be an efficient way to both pull and push data from and to Databricks.

Finally, we tried a few different techniques to pull data out of Databricks for external consumption. Overall, we got an insight in how data can flow into and out of Databricks using a few simple commands. In the next chapter, we'll start coding for real, using SQL.

CHAPTER 6

Querying Data Using SQL

Finally! We have our data loaded and ready in Databricks – multiple exciting datasets to investigate. Now it's time to start playing around with them. We'll start by using one of the oldest data languages around.

In this chapter, we'll look at Spark SQL and how it works in Databricks. We'll pull data from tables, filter it according to our needs, and look at aggregation functions to investigate our datasets. We'll also look at how Delta Lake pushes Spark SQL and Databricks closer to the classic database feature set and what that means for manipulating data in place.

As you will see over the next few chapters, there are many ways to look at, process, and create data in Databricks. Many changes have come and gone over the years, but for traditional data analytics, it is still the almost 50-year-old SQL, an acronym for Structured Query Language, that dominates.

No matter what other tools you use, you pretty much have to know the basics (and preferably a bit more) of SQL. If you have ever accessed data from any type of relational database, it's very likely that you already have had exposure to it.

SQL is a very good tool for quickly getting a feel for the dataset you are working with. Even if you do most of your processing in, for instance, Python, it's in many cases faster to run a few queries in SQL to see what the data looks like.

The Databricks flavor

Although SQL is used in many products and is similar across the board, there are differences in different implementations. A lot of syntax that works in Oracle RDBMS won't work in Microsoft SQL Server and vice versa. There is an ANSI standard that companies adhere to somewhat, but most tools contain features that are not available in the competitors' products.

It's the same for Spark SQL in Databricks. Most of the core statements work as you'd expect, but there are differences compared to the major players in the field. For many features you'd consider standard, such as UPDATE, you also need to use the fairly new Delta Lake component.

As this is not primarily a book about SQL (there are many good ones out there if you're new to it), we'll not dig too deeply into the nuts and bolts of the language. Still, we'll spend a little bit of time on the basics in this chapter. It's important for you to understand how to use it and what Databricks flavor looks like, especially as SQL for many is the be-all and end-all even in Databricks. Just being able to run traditional queries on huge datasets is good enough for a large number of analysts.

Let's get going!

Getting started

The first thing you have to do is to start a cluster, as described in Chapter 5. Then create a notebook with SQL as the primary language. Connect to the cluster you started and wait for the notebook to open.

Before we start running our queries, we need to let Databricks know which database we want to work with. The normal behavior is that every command you execute will be run on the "default" database.

As I mentioned in earlier chapters, it's usually not a good idea to store tables in the default database. Everything we just loaded is neatly stored in their own databases. So we'll have to let Databricks know where our data is stored. There are a few ways to do this. The easiest way is to simply type in USE, followed by the database name.

Before we can do that, we need to get the data. In this chapter, we'll spend a bit of time using the New York Taxi dataset. You can pick it up at `www1.nyc.gov/site/tlc/about/tlc-trip-record-data.page`. I'll be using the Yellow Taxi Trip Records data for June 2019 and the Taxi Zone Lookup Table. Use the same if you want all the following commands to just work.

Before you import the downloaded files, create a database where you can store them. In the SQL notebook, run the following command. Once you've run it, verify that you can see it if you click the Data button:

`CREATE DATABASE taxidata`

CHAPTER 6 QUERYING DATA USING SQL

Once this is done, just import the data you just downloaded. You can use the graphical interface for this. Make sure it ends up in the taxidata database, you use the first row for headers, infer schemas, and you call the tables yellowcab_tripdata_2019_06 and taxi_zone_lookup, respectively.

As we'll in this chapter use this dataset, let's point to that. All the commands we execute following this point will be run on the taxidata database – very convenient if you want to just work with one dataset, like in our case:

USE taxidata

Using this alternative can be a bit cumbersome if you want to work with multiple databases at the same time though. It's even impossible to use if you need tables from two databases in the same query. Luckily you can instead prefix the table name with the database name, like this:

SELECT * FROM taxidata.yellowcab_tripdata_2019_06;

This will pick up data from the taxidata database no matter what the current active database is. If you are uncertain of which database is currently activated, you can check by running the following command:

SELECT current_database()

In our case, we'll just use data from one database. So all the commands in this chapter expect you to have run the USE command from the preceding text. If you haven't already, please go ahead and do so, or you'll get a lot of errors.

Picking up data

As you've seen in the last two examples, looking at data is as easy as typing SELECT. At its core, it's a rather simple command. You type in that you want to get data, define what columns you're interested in, and then finalize the command with the table name:

SELECT * FROM yellowcab_tripdata_2019_06;

This returns the first 1,000 rows in a grid. The asterisk is shorthand for all columns in the table. That means you can type the * sign instead of vendorid, tpep_pickup_datetime, tpep_dropoff_datetime, and so on. Convenient. Just be aware that getting data back is often faster if you limit the output to the columns you are interested in as

Databricks has to process less data. Also, it's more obvious for future readers of your code to follow what's happening:

```
SELECT VendorID, Trip_Distance FROM yellowcab_tripdata_2019_06;
```

This will only give you the data from VendorID and Trip_Distance, instead of everything in the table. If it's a wide table with a lot of columns and a very large number of rows, the difference in performance can in some operations be substantial. It's not so much if the only thing you want to do is look at the first rows though as that's pretty much always fast.

If you want to limit the number of rows shown returned, you can use the LIMIT keyword. This is an efficient way to just get the top rows from a table and is usually used in combination with filters, which we'll get to in a bit. This command will give you the first ten rows:

```
SELECT VendorID, Trip_Distance FROM yellowcab_tripdata_2019_06 LIMIT 10;
```

By the way, did you notice the semicolons at the end of the commands? When you only have one SQL command in a cell and execute it, it'll run just fine even if you don't have a semicolon. But if you run multiple commands together, Databricks won't know how to separate them unless you use a semicolon. Try running this, for instance:

```
SELECT Trip_Distance FROM yellowcab_tripdata_2019_06
SELECT VendorID FROM yellowcab_tripdata_2019_06
```

This will create an error. The reason is that you can run statements over multiple lines. Line break doesn't really mean anything in SQL. To avoid confusing the parser, get into the habit to always add the semicolon to your queries. It'll save you some headache in the future.

Long names are boring to type and cumbersome to read. Luckily SQL has support for aliases, which makes your life easier and your code clearer. You basically give your tables and columns an additional name to reference. Note that the alias only works for the current query and not for subsequent ones:

```
SELECT tpep_pickup_datetime as pickup, tpep_dropoff_datetime dropoff FROM yellowcab_tripdata_2019_06 as T
```

As you can see in the preceding example, it is ok to use the AS command before the alias, but it's not necessary. Aliases like these are possible to reference later in the query, as you will see when we get into joining data. In those cases, they help in readability a lot.

A word of caution: It's easy to use shorthand, like the preceding T. But when you come back to the code later on, it's frustrating to wade through complex SQL referencing T1, T2, T3, and so on. Try using aliases that make sense.

Filtering data

Ok, let's continue. In most cases, you're not just interested in getting all the rows just straight up. Beyond a few hundred rows, it's very hard to make any sense of information in a spreadsheet kind of way. One way to cut down the information is to filter it using the WHERE clause:

```
SELECT * FROM taxi_zone_lookup WHERE borough = 'Queens';
```

Now you only get a subset of the data. More specifically, you only get the rows where the column *borough* matches the term *Queens* exactly. Note that the string is case sensitive, so you wouldn't get any rows where *borough* equals *queens,* for instance.

This is not the only thing you can do with the WHERE clause. There are actually quite a few ways of filtering data to the subset you want. If you, for instance, have a list of items you want to match, you can run the query with the IN clause:

```
SELECT * FROM yellowcab_tripdata_2019_06 WHERE VendorID IN (1,2);
```

In many cases, you don't know exactly the term you are looking for, but only part of the name. Or maybe the number of items you want is too long to type out in a long IN list. In that case, LIKE might be your solution:

```
SELECT * FROM taxi_zone_lookup WHERE borough LIKE 'Staten%';
```

Here % is a wildcard character. So your query is looking for strings starting with the string *Staten*. This query will match *Staten Island* and *Staten Mainland* (if such a place existed) but not *Mystaten*. Luckily you can use multiple wildcards and put them pretty much wherever you want in the string:

```
SELECT * FROM taxi_zone_lookup WHERE borough LIKE '%een%';
SELECT* FROM taxi_zone_lookup WHERE borough LIKE 'M%nha%n';
```

Let's try another common way of filtering data: ranges. If you want to get all values between 50 and 100, you could create a list, but it'd be cumbersome. Instead, you probably want to use BETWEEN. So you get

```
SELECT * FROM yellowcab_tripdata_2019_06 WHERE vendorid BETWEEN 1 AND 5;
```

This works well for dates as well in case you want to filter out data for a given period of time. As you might imagine, that is pretty common in data analysis scenarios. The syntax is the same, but with dates in the yyyy-MM-dd format:

```
SELECT * FROM yellowcab_tripdata_2019_06 WHERE tpep_pickup_datetime BETWEEN '2019-06-02' AND '2019-06-03';
```

If you are looking at those dates and thinking that the format is a strange one, don't worry. You can use functions to use the formats you are used to (although you really should start using the ISO standard for dates). I'll explain how functions work in a bit, but here's a little taste:

```
SELECT * FROM yellowcab_tripdata_2019_06 WHERE tpep_pickup_datetime BETWEEN to_date('06/02/2019','MM/dd/yyyy') AND  to_date('2019/06/03','yyyy/MM/dd');
```

Even though you don't need an explicit call to the functions if you use the ISO standard format, it's still a good idea to have it. That said, the actual query does get a bit more cumbersome to read and understand with all the extra function calls.

There is another way to filter data that you'll probably use a lot. Spark SQL, like most databases, supports something called subqueries. That means that you use the result of one query as input to another one:

```
SELECT * FROM yellowcab_tripdata_2019_06 WHERE pulocationid IN (SELECT locationid FROM taxi_zone_lookup WHERE zone = 'Flatlands')
```

This inner SELECT will return the locationid value for the zone called *Flatlands*. The outer SELECT will use the result of this query to filter out all rows where the pickup location is in the Flatlands.

Subselects are a powerful way of filtering data based on another table. You might have a long list of items, for instance, and want to match the rows in the sales table for a particular group. This is a quick way of doing lookups in dimension tables, among other things.

Also worth mentioning is that you can use NOT in combination with the preceding commands. So NOT IN and NOT LIKE work just fine. Just be aware that NOT is frequently computationally more expensive than normal matching:

```
SELECT * FROM yellowcab_tripdata_2019_06 WHERE vendorid NOT BETWEEN 1 AND 5;
```

You can also use the logical operators AND and OR in your WHERE clauses to combine multiple filters. They can be combined with nearly any other operator. Just be careful when using OR in more complex queries. Use parentheses to make sure you get what you want. Consider the following:

```
SELECT * FROM taxi_zone_lookup WHERE borough = 'Queens' OR (borough = 'Staten Island' AND zone = 'Arrochar/Fort Wadsworth');
SELECT * FROM taxi_zone_lookup WHERE (borough = 'Queens' OR borough = 'Staten Island') AND zone = 'Arrochar/Fort Wadsworth';
SELECT * FROM taxi_zone_lookup WHERE borough = 'Queens' OR borough = 'Staten Island' AND zone = 'Arrochar/Fort Wadsworth';
```

The first one will give you all rows related to Queens and the Arrochar zone in Staten Island. The second one will only give you the Arrochar row. Can you guess what result set you will get back with the last example? It's very easy to make a mistake, so try to always clearly encapsulate your OR statements to avoid confusion and hard-to-find bugs.

Now, there is one more thing worth mentioning here. That is the concept of NULL. It literally is missing information. That means that you can run queries such as the following ones and not get all the rows in the dataset:

```
CREATE TABLE nulltest (a STRING, b INT);
INSERT INTO nulltest VALUES ('row1', 1);
INSERT INTO nulltest VALUES ('row2', NULL)

SELECT * FROM nulltest WHERE b = 1;
SELECT * FROM nulltest WHERE b != 1;
```

Logically a value is either 1 or not 1, so you should get all rows, right? Not so if your column contains a NULL value. As the content of the cell is missing, it'll never turn up in a normal query. Instead you have to specifically look for them:

```
SELECT * FROM nulltest WHERE b IS NULL;
SELECT * FROM nulltest WHERE b IS NOT NULL;
```

It's possible to block NULLs from ever entering your table by creating constraints on the schema, but it's not uncommon to get these non-values in data. You will also create them yourself when you join data. We'll come across them many times in this book, and you'll see why it can happen. For now, just be aware that they exist. They are a common source of mistakes.

Joins and merges

In many cases, you'll not work with a single table or even a single schema. Instead you need to combine data from multiple sources. Getting to a point where this is possible is a major part of the data engineers' and data scientists' workload, and we'll talk more about that in the Chapter 8. Once you have the data organized and connectable however, it's easy to run the actual operations.

If you have multiple tables with the same schema, or at least one common column, that you just want to put on top of each other, you use a UNION statement. For instance, you might have sales data for two different years and want to create one dataset of them:

```
SELECT * FROM yellowcab_tripdata_2019_05
UNION
SELECT * FROM yellowcab_tripdata_2019_06;
```

For this to work, the table schemas need to match perfectly. If the tables only partially match up, you can specify the columns, and it'll work just fine. If you know there are no duplicates (or if you want them) in the dataset, use UNION ALL, as that will skip duplicate checks and just slam the data together.

> **THE DIFFERENCE BETWEEN UNION AND UNION ALL**
>
> While I mentioned the difference between UNION and UNION ALL in passing, it's worth mentioning again as I do see there's a misunderstanding about what they do. This is partially due to the fact that it works differently in Python.
>
> The UNION ALL command will just combine the datasets without doing any duplication checks on the data. To see the difference, try to run the following commands and look at the output.

```
create table x (a integer);
create table y (a integer);
insert into x values (1);
insert into x values (2);
insert into y values (1);
insert into y values (2);
insert into y values (3);
select * from x union select * from y;
select * from x union all select * from y;
```

As you can see, you'll get duplicates in the latter case. If you can live with this or know for a fact there aren't any in the tables you're using, you should always go with UNION ALL. It's faster as Apache Spark doesn't have to do an expensive check for identical rows. The difference for large datasets can be huge.

But wait! There's more. The INTERSECT command will create a result set with data that exists in both tables, basically a list of all duplicates. With MINUS, you get everything from table1, except for the duplicates. They are removed.

In most cases, you want to connect datasets that are not identical, but rather just share some attribute. Maybe you have a list of customers in one table and all sales transactions in another. To connect sales to users, you need to use a join. It basically connects data from two or more (usually normalized) tables. If you're not familiar with relational modeling, check the appendix for a short introduction:

```
SELECT
  tz.Borough,
  tz.Zone,
  yt.tpep_pickup_datetime,
  yt.tpep_dropoff_datetime
FROM
  yellowcab_tripdata_2019_06 yt
  LEFT JOIN taxi_zone_lookup tz
  ON (yt.PULocationID = tz.LocationID);
```

This will give you the borough and zone from one table and the dates from the other, in one query. That is pretty awesome. The trick is that the pulocationid and locationid columns match up.

In some cases, joins can create those pesky null values we talked about earlier. Consider a borough with no fares, for instance. The two first columns will be populated, but the last two (pickup and drop-off times) won't. Instead they turn into nulls.

Sometimes you want these values, sometimes you don't. Maybe you want to show all boroughs, no matter if there are any rides or not. Maybe you don't. Although you could remove those rows later with a filter, it's smarter to decide if you want those rows or not in advance. You do that with different join statements:

```
SELECT c.customer_name, SUM(t.sales) FROM customer c INNER JOIN transactions t ON (c.cust_id = t.cust_id);
SELECT c.customer_name, SUM(t.sales) FROM customer c LEFT JOIN transactions t ON (c.cust_id = t.cust_id);
SELECT c.customer_name, SUM(t.sales) FROM customer c RIGHT JOIN transactions t ON (c.cust_id = t.cust_id);
```

The order of statements here matters. INNER, which is the default if you just type JOIN, will only match data that is available on both sides. LEFT will show all the rows from the table to the left of the command and only matching data from the right side. RIGHT is just the opposite.

As mentioned, you don't have to limit yourself to two tables. Just remember that the more you add, the bigger the chance the optimizer chokes on the query and creates a bad execution plan, resulting in a slow query.

Ordering data

Normally you want to sort the dataset you get back. At least if it's meant for human consumption. The way to do this in SQL is by using ORDER BY. And you basically just list which columns, in order, you want to sort by. If you want to primarily order by column 1 and secondarily by column 2, you write

```
SELECT * FROM taxi_zone_lookup ORDER BY borough, zone;
```

If you want reversed sorting, just add DESC (descending) after the column. The opposite, ASC (ascending), is implicitly used by default. It's ok to mix and match both versions in the same statement:

```
SELECT * FROM taxi_zone_lookup ORDER BY borough DESC, zone ASC;
```

This would first order the data by borough in a descending order and then order it by zone in an ascending order. As I just mentioned, you don't really need to type out ASC, and you'll see that it's rarely done.

Functions

Once you've looked a bit on the data, you might want to do some processing, like summing up some values. This is where the real analytics work starts. Luckily there's a useful set of aggregate functions available to you in SQL. You can group information and then let Spark count, summarize, or run other mathematical algorithms or functions on the data, for instance:

```
SELECT ROUND(SUM(trip_distance),2) FROM yellowcab_tripdata_2019_06;
SELECT VendorID, SUM(fare_amount) total_amount FROM yellowcab_tripdata_2019_06 GROUP BY VendorID ORDER BY total_amount;
```

The first example returns the total number of miles driven during the month of June. Note, by the way, how we stack functions on top of each other. We use ROUND on top of SUM to get a readable number with just two numbers to the right of the decimal point.

Next is a list of vendors and their total fare amount for June. Those are some big numbers. There seems to be more than a few dollars in the taxi business! Anyways, we use aliases to simplify sorting. As mentioned before, the default ordering is ascending, so the biggest earner is at the bottom.

Note that all the columns that are not aggregated have to be in the GROUP BY clause in the end. Otherwise, you'll get an error message. It can sometimes be easy to forget this, and the error you get back is not always clear.

There are many other functions you can do as well. You can easily pick up the lowest value, the highest, and the average. It's also possible to filter the result set in the same query, using the HAVING clause:

```
SELECT MIN(fare_amount), MAX(fare_amount), AVG(fare_amount) FROM yellowcab_tripdata_2019_06;
SELECT PULocationID, MIN(fare_amount), MAX(fare_amount), AVG(fare_amount), COUNT(fare_amount) FROM yellowcab_tripdata_2019_06 GROUP BY PULocationID HAVING COUNT(fare_amount) > 200000;
```

CHAPTER 6 QUERYING DATA USING SQL

Not bad. Already with this very limited part of SQL, you can do a lot of investigation and get a feel for the dataset you're looking at. Now, as you might remember, we can also get a graphical view of the data. With the last query in the view, click the bar chart button and then Plot Options. Add AVG(fare_amount) to the Values box and remove the count field. Now you have a nice little view of the dataset that clearly shows an outlier.

On top of all the math functions, there are tons of other functions built in, literally hundreds. We've already seen the date functions earlier in this chapter, and you'll see more later. Most of them are related to math, dates, or strings, but not all. For instance, you can use md5 to get an MD5 128-bit checksum:

SELECT md5 ('Advanced Analytics')

Note that we didn't specify a table. A lot of functions will work in this way. You can just type SELECT, the function name, and the parameters. Databricks will automatically assume you have a magic row. Normally you'd want to run the function on a table column though.

Windowing functions

There are another set of powerful analytics tools in SQL that we should look at more closely. They are called windowing functions and are very useful in analytics. They offer you an ability to work with functions on a set of rows. This, for instance, makes it possible to rank data, compare rows to each other, and get both partial and total aggregates in the same result set – plus a lot more.

We can try this out by looking at the fares for each day and at the same time checking the total – a tricky thing to do that requires multiple queries without our windowing functions. Much easier with this new toolset. Let's start with creating an aggregated table as base:

CREATE TABLE taxi_june_day_sum AS SELECT dayofmonth(tpep_pickup_datetime) day, passenger_count, sum(fare_amount) total_fare_amount FROM yellowcab_tripdata_2019_06 GROUP BY dayofmonth(tpep_pickup_datetime), passenger_count;

This will give you an aggregated table with the total fare amount grouped on the number of passengers and the day of the month – an interesting analysis in itself. Now we need to add the total and compare the day fares to it:

```
SELECT
 day,
 passenger_count,
 total_fare_amount,
 round(sum(total_fare_amount) OVER (PARTITION BY passenger_count),2)
 passenger_total,
 round(total_fare_amount/sum(total_fare_amount) OVER (PARTITION BY
 passenger_count) * 100,2) pct
FROM
 taxi_june_day_sum;
```

The query might be a bit cumbersome to parse through, but take it step by step and you will understand it. We get the day, the passenger count, and the fare amount. Then we use the SUM-OVER-PARTITION BY construct to sum over passenger count. Following that, we do the same thing again, but this time we divide it with total fare amount to get the percentage of the total.

Another case where traditional SQL has a hard time is with accumulation. If you want to increase the row value into a variable and show it continuously, you've historically been out of luck. Again window functions come to the rescue:

```
SELECT
 day,
 passenger_count,
 total_fare_amount,
 sum(total_fare_amount) OVER (ORDER BY day, passenger_count ROWS BETWEEN
 UNBOUNDED PRECEDING AND CURRENT ROW)
FROM
 taxi_june_day_sum
ORDER BY
 day,
 passenger_count;
```

Unlike the previous example, we run a separate aggregation command on the complete dataset, but we sort it. That will give you the running total and also the buildup toward the total. On the last row, you'll have the same value as if you just sum the column.

These are but a few examples of how you can use windowing functions. Databricks supports many more: Rank, Dense_Rank, Percent_Rank, Ntile, and Row_Number for ranking and Cume_Dist, First_Value, Last_Value, Lag, and Lead for analytics. On top of this, you can use many of the normal functions, like sum and avg, as you've seen.

Here's another example with LAG. This picks the value from a former row and displays it alongside the current one. With this you can, for instance, calculate the time between the current row and the last one. The second argument to LAG is the number of rows you want to go back:

```
SELECT
 vendorid,
 tpep_pickup_datetime,
 LAG(tpep_pickup_datetime,1) OVER (ORDER BY tpep_pickup_datetime) lag1
FROM
 yellowcab_tripdata_2019_06;
```

Even with windowing functions, it might be hard to get everything you want to do into one statement. If you need to run multiple operations and don't want to save the result set in between, you can in some cases use subselects:

```
SELECT * FROM (SELECT day, total_fare_amount, dense_rank() OVER (ORDER BY total_fare_amount desc) ranking FROM (SELECT day, sum(total_fare_amount) total_fare_amount FROM taxi_june_day_sum GROUP BY day)) WHERE ranking <= 10 ORDER BY ranking;
```

This will give you the list of the ten days with highest total sales. The trick is that you select from the result of another select. Twice! You can do this in many layers and in parallel but beware – your SQL will be hard to read. You can use WITH to clean it up a bit:

```
WITH q_table AS
(SELECT day, sum(total_fare_amount) total_fare_amount FROM taxi_june_day_sum GROUP BY day)
SELECT * FROM (SELECT day, total_fare_amount, dense_rank() OVER (ORDER BY total_fare_amount desc) ranking FROM q_table) WHERE ranking <= 10 ORDER BY ranking;
```

CHAPTER 6　QUERYING DATA USING SQL

A note about rank, since I used it in the preceding code. There's also a windowing function called dense_ranked. The difference is that rank will skip numbers after a tie, while dense_rank will keep all numbers. So rank might give you 1, 1, 3, 4 and dense_rank 1, 1, 2, 3 for the same dataset.

As you've seen, there are a large number of functions in Databricks; and if you want even more, you can create them yourself. Just remember that functions can be costly and using them slows things down. This is especially true for row-based evaluations. So if you don't need to use things like UPPER, don't. Preparing data in a consistent manner is usually better.

A view worth keeping

Frequently you end up with queries that you want to run over and over. If it's a complicated join, it can be cumbersome to type it in every time. The solution is to use a view, which is basically a stored query with a name attached to it:

```
CREATE VIEW borough_timespan_view AS SELECT tz.Borough, MIN(yt.tpep_pickup_datetime) first_ride, MAX(yt.tpep_pickup_datetime) last_ride FROM yellowcab_tripdata_2019_06 yt LEFT JOIN taxi_zone_lookup tz ON (yt.PULocationID = tz.LocationID) GROUP BY tz.Borough;
```

Now that you have your view, you can query it just like any table. You can even join with other tables and add WHERE clauses for further filtering. Under the surface, Databricks will recreate the query and execute it in the best way possible, just like if you typed it in:

```
SELECT * FROM borough_timespan_view;
```

Views are also frequently used to expose data to external users and applications. By adding a logical layer, you can change the underlying structure without having to alter applications, some of which you might not have control over.

They can also help you hide data that you don't want others to see. By only showing some columns of a table, you can provide data while hiding parts of it. We'll talk more about this in Chapter 11 where security will be brought up.

One more thing about views: When you create them as in the preceding text, they'll by default be available for other users. If you only want them to exist in your current notebook, you can add the TEMPORARY keyword. You can also add REPLACE to automatically drop and recreate a view:

```
CREATE OR REPLACE TEMPORARY VIEW number_of_rows_view AS SELECT count(*) FROM yellowcab_tripdata_2019_06;
```

This will create a view (and replace it if it already exists) that you can use in the current notebook for the duration of your session. You'll primarily need this when you are writing cleaning jobs. It's frequently helpful to have supporting tables when running large notebooks, but you don't always want to materialize them.

Hierarchical data

So far we've only talked about traditional tables with rows and columns. Nowadays you'll frequently come across more complex structures even in text files. Most notably, you'll get data in the JavaScript Object Notation, of JSON, format.

SQL isn't really built for this types of structure, but it works pretty well in most database products. In Spark SQL, and therefore Databricks, the support is exceptional; and while it's still easier to work with JSON using Python or R, you can do it in plain SQL as well. Let's start by creating a table with JSON data:

```
%python
import json

x = '[{"brand":"Apple"
            , "models":["MacBook Air","MacBook Pro"]}
        ,{"brand":"Dell"
            , "models":["XPS","Latitude"]} ]'
js = json.loads(x)
with open('/dbfs/tmp/json_example.json', 'w') as outfile:
    json.dump(js, outfile)
```

We create a JSON file using Python. The data just contains a couple of computer brands with two models each. The loads and dump commands make sure what we store is actually parsed as JSON. Any formatting errors will throw an exception:

```
CREATE TABLE json_example USING json OPTIONS (path "/tmp/json_example.json", inferSchema "true");
```

This will load the example file with JSON data into a new table. If you just query it like a normal table, you'll notice that some of the column output doesn't look like what you're used to. In this case, models is an array with several items on the same row. You'll also see a small arrow with which you can expand or collapse the data:

```
%sql
SELECT brand, models FROM json_example;
```

If you want to look at the data in a more traditional way, you have the option to explode the array into rows. This is not just a good way to get a better view of the data but also simplifies filtering:

```
%sql
SELECT brand, explode(models) FROM json_example;
SELECT * FROM (select brand, explode(models) as model FROM json_example) WHERE model = 'MacBook Pro';
```

Even if you might not store data in JSON format in your own database, this is a pretty good feature to have when you are pulling in data from other sources. You can quickly get a feel for the structure and see what you ultimately want to store in normalized form.

Creating data

So far we've mostly been selecting data from existing tables or files. It's time to create a few tables from scratch and fill them with new data. We do that using Data Definition Language commands, or DDL for short.

Creating a table is easy, as you've already seen here and in Chapter 5. You either need to have a dataset you can use as the schema or define what the table should look like manually. If you, for instance, want to store a list of taxi drivers, you can create

```
CREATE TABLE taxi_drivers (taxi_driver_id BIGINT NOT NULL, first_name STRING, last_name STRING);
```

Here we basically just created a table with everything set to default, with the exception of banning NULL values from taxi_driver_id. Not that Spark SQL does not uphold constraints like this. You can still insert NULL values into the first column. The information is mainly used by the optimizer to make smarter decisions on how to pick up data. Beware of this little issue.

Using the default settings might or might not be sensible. For a small table like this one, defaults will be just fine. Larger ones might benefit from handling partitions and clusters as shown in Chapter 5.

If you run this command again, you'll get an error. While that doesn't matter much if you work manually step-by-step, it's not great in a script. It's better to tell Databricks to just ignore the create statement if the table already exists:

```
CREATE TABLE IF NOT EXISTS taxi_drivers
(taxi_driver_id BIGINT NOT NULL
,first_name STRING
,last_name STRING);
```

One thing that is usually a good idea to do, and far too many don't, is to add comments explaining what we just built. You can also use table properties to define your own tags and values. Future users of the table (including the older version of yourself) will thank you if you do this:

```
CREATE TABLE taxi_drivers
(taxi_driver_id BIGINT COMMENT 'This is a generated key'
,first_name STRING
,last_name STRING)
COMMENT 'This contains all the taxi drivers driving in NY' tblproperties
('created_by'='Robert');
```

If you in the future want to see the comments and table properties for this table, or any other for that matter, you just use the DESCRIBE command, or DESC for short. You just have to add the EXTENDED parameter:

```
DESC EXTENDED taxi_drivers
```

Scroll down, and among a lot of information, you'll find both Comment and Table Properties, with the values displayed. If you find a new table and if you are lucky, you might get some insight this way. This is even truer when it comes to columns.

If you need to change anything in your table after the fact, there's an ALTER command to help you out. With this command, you can add or change columns without having to recreate the table:

```
ALTER TABLE taxi_drivers ADD COLUMNS (start_date TIMESTAMP COMMENT 'First day of driving' AFTER taxi_driver_id);
```

So far we only created the structure of the table. It's still totally empty. Let's get a few rows in, just to test it out. Once you entered the data, feel free to look at it so you know for sure it's there:

```
INSERT INTO taxi_drivers VALUES (1, current_date(), 'John', 'Doe');
INSERT INTO taxi_drivers VALUES (2, NULL, 'Jane', 'Doe');
SELECT * FROM taxi_drivers;
```

In the second example, we use NULL. We just didn't know when Jane started, and as the column is a timestamp field, we can't just type out unknown. Databricks will expect you to type in all values, or you'll get an error.

A somewhat strange omission from Spark SQL is that you can't specify what columns you want to insert data to. In most databases, you can write something on the lines of the following statement, but this will return an error in Databricks:

```
INSERT INTO taxi_drivers (taxi_driver_id, first_name) VALUES (3, 'Ronda');
```

Once you're done with all this work, it's time to get rid of it. You don't want the European Union smacking you down with the GDPR hammer, so a table with identifiable names has to go. Luckily it's easily done:

```
DROP TABLE taxi_drivers;
```

It's good to know how to create tables manually as you've seen here and in the last chapter. But in data analysis, it's more common to make them using another tables or files as the template. If you only want to work with data related to rides with multiple passengers, it makes sense to create a new table based on that data. That way you'll read less data every time you work with your dataset:

```
CREATE TABLE multiple_passengers AS SELECT * FROM yellowcab_tripdata_2019_06 WHERE passenger_count > 1;
```

The new table will inherit the table structure of the original and automatically be populated with the data returned from your query. The SELECT statement can be however complex you want. Just remember to set aliases for aggregated columns, or it won't work.

Manipulating data

In the preceding text, we used the INSERT command to add new data to a table. This is an operation within a group of functions within the Data Manipulation Language, or DML. Whenever you want to create a new object and fill it up within Databricks, this is what you'll use – implicitly or explicitly.

The most basic use is to manually add a handful of rows, as we've seen. You just have to tell Databricks which table you want to take the data from and then define it, according to the schema.

Manually adding data in this way is not very common though. It's much more frequent to add data from another source. Either you create a new table based on a query as shown earlier, or you just add data to an existing table:

```
CREATE TABLE yellowcab_tripdata_pass_part AS SELECT * FROM yellowcab_tripdata_2019_06;
INSERT INTO yellowcab_tripdata_pass_part SELECT * FROM yellowcab_tripdata_2019_06 WHERE passenger_count = 3; ;
```

Oops, we just duplicated data where the number of passengers was 3. If you want to replace data, the traditional way of doing it is to first wipe out the information using TRUNCATE TABLE. This will quickly get rid of all the data you have in a table or a given partition. In our case, we'll use the second option:

```
TRUNCATE TABLE yellowcab_tripdata_pass_part;
TRUNCATE TABLE yellowcab_tripdata_pass_part PARTITION (passenger_count=3);
```

A nice feature with Databricks implementation of INSERT is that you can do the cleanup in the same command. This makes reloading tables even easier, which is always a good thing:

```
INSERT OVERWRITE TABLE yellowcab_tripdata_pass_part SELECT * FROM yellowcab_tripdata_2019_06;
```

You can also use INSERT to save data into a directory. This make sense if you, for instance, use Databricks to populate an external storage with data or just want to get some data out quickly in a general format:

```
INSERT OVERWRITE DIRECTORY '/mnt/export/json/taxidata' USING json SELECT *
FROM taxi_zone_lookup;
```

This assumes you have the folder structure mounted to the preceding path, which you hopefully did in Chapter 5. The result is a json file with the data from taxi_zone_lookup inside the taxidata folder, together with some metadata files.

I should mention that you can do this exact thing in many ways. You can use Python, as you'll see in the next chapter, or even the CREATE TABLE as it's actually possible to store a table as a text file.

Delta Lake SQL

As I mentioned in Chapter 2, data stored in RDD is immutable to make it scalable and fast. That is, however, a big problem in many common use cases. Having to rewrite large tables to do a few updates can be frustrating and time consuming.

In Chapter 3, we discussed how Databricks is moving into traditional database territory with the introduction of Delta Lakes. This might not seem like a big deal, but it's actually huge as it makes Databricks more usable for traditional business intelligence methods.

It specifically simplifies things for you when you are working with cleaning the data. Reading the pros in the Databricks press release for the general availability of the feature is a little bit funny as they kind of rediscovered databases. The problems they see are as follows:

1. Reading and writing to data lakes is not reliable.

2. The data quality in data lakes is low.

3. Poor performance with increasing amounts of data.

4. Updating records in data lakes is hard.

So a big, steaming pile of unstructured data is apparently hard to manage and use efficiently. Who would have guessed? Sarcasm aside, it's a good thing they decided to tackle this issue though as the benefits are many as previously noted.

From a SQL perspective, it means that you get access to a lot more data manipulation options – more specifically UPDATE, DELETE, and MERGE. These commands will let you alter data in place. They are (together with INSERT) part of the DML set of operations.

This is the bread and butter of traditional, operational databases. In decision support systems, they are generally avoided when possible as the operations are pretty expensive. Having access to them is a big benefit, but think hard about if you really need them.

When you create a table, you need to explicitly specify that you want to use it as a Delta Lake table. Let's create a delta version of our taxi zone table. As we'll fill it with the same data, we can use the old table as a template. Note that we use USING DELTA when creating the table:

```
CREATE TABLE tzl_delta USING DELTA AS SELECT LocationID, Borough, Zone, service_zone FROM taxi_zone_lookup;
```

Now we have a table on which we can run DML commands. On the surface, you won't notice any difference as the tables are compliant with most features in Databricks. You'll just be able to do more. Let's see what options you have.

UPDATE, DELETE, and MERGE

First up is UPDATE. This you can use if you have some pesky little error found in your data and want to correct it. Just let Databricks know what table to fix, what the fix is, and the filter condition:

```
UPDATE tzl_delta SET zone = 'Unknown' WHERE locationid = 265;
```

This will change the row with locationid 265 so that zone is changed from NA to Unknown. You can of course use the WHERE clause to change multiple rows at the same time. This will change the zones of all unknown boroughs to Unknown:

```
UPDATE tzl_delta SET zone = 'Unknown' WHERE borough ='Unknown';
```

If you don't add any WHERE clause, you'll update all rows. This is normally a bad idea unless it's a table with few rows as the UPDATE command is slow. In most cases, it's actually faster to rewrite the whole table, or partition, with the correct data even if only a minor part of the data needs to be changed.

CHAPTER 6   QUERYING DATA USING SQL

Another tool you probably want to use sparsely is DELETE. As you can probably guess, this is the command to remove unwanted data. Just like with the UPDATE command, you filter using WITH:

```
DELETE FROM tzl_delta WHERE locationid = 265;
```

The result, as you might expect, is that all rows with the value in the column locationid being 265 will be removed from the dataset. Again, this is a costly operation; and if you could avoid it, you should.

One command that you probably will want to use more frequently is MERGE. This little command will make your life much easier when you transform data. It basically lets you combine data while using a rule set. You can UPDATE, INSERT, and DELETE – all in one statement. Let's prepare an update table:

```
CREATE TABLE taxi_zone_update AS SELECT * FROM taxi_zone_lookup where 1=0;
INSERT INTO taxi_zone_update VALUES (264, 'Unknown', 'Not applicable', 'Not applicable');
INSERT INTO taxi_zone_update VALUES (265, 'Upcoming', 'Not applicable', 'Not applicable');
```

The first row creates a copy of taxi_zone_lookup with no data as the predicate is never true. Then we insert a couple of rows to handle the unknown rows in the main dataset. Now let's update the old dataset with this new information:

```
MERGE INTO tzl_delta tz
USING taxi_zone_update tzUpdate
ON tz.locationID = tzUpdate.locationID
WHEN MATCHED THEN
  UPDATE SET borough = tzUpdate.borough, zone = tzUpdate.zone, service_zone
  = tzUpdate.service_zone
WHEN NOT MATCHED
  THEN INSERT (locationid, borough, zone, service_zone) VALUES (tzUpdate.
  locationid, tzUpdate.borough, tzUpdate.zone, tzUpdate.service_zone);
```

That might look a bit scary at first, but it's really not. Let's go through it. We start with defining the target table, tzl_delta. Then we pick the source and define the join condition to use. Then comes the logic: one row for UPDATE and one for INSERT. We could have added a DELETE as well.

This is frequently the best command for continuously adding new data to an existing dataset, like daily loads. If you track logistics, for instance, the shipment status will change over time and you might want to change data based on some rule set.

An interesting note is that it's in many cases faster to use MERGE instead of UPDATE even if you just want to update information. This is especially true if your changes are dependent on a complex join.

Keeping Delta Lake in order

One problem with Delta Lake is that you might lose performance over time. There are a few ways you can try to mitigate this. The first one is to run an optimization command. We'll go through it in the optimization discussion, but already now you can test running

```
OPTIMIZE tzl_delta
```

Interestingly enough you have the option to only run this process for a small subset of the data. So if you can, for instance, choose to compact data you know you'll be using soon and leave the historical data be or process it at a later time.

Either way this work happens in the background and won't affect other users querying the data. Still, it does consume a lot of resources, so you don't want to do it unnecessarily. Strangely enough it's a manual process. There is no way to ask Databricks to do it automatically.

Although they do claim there are good reasons for this, like resource usage and not knowing how you use the tables, they could – and should – give you the option to have some automation if you want it. Hopefully you'll see this in the future. For now you'll have to schedule it (see Chapter 11 for more about scheduling) or add it to your jobs.

So how frequently should you run this command? There is unfortunately no hard and fast rule you can follow (if there were, Databricks would use it and automate accordingly). As the result of the operation is faster access times, you need to weigh time to compress to time spent querying the data. It makes no sense to do this for data you will only read once, for instance. If you have 100 jobs using the table though, it makes a lot of sense to run one big job to speed up the 100 following ones.

Another thing you might want to look at is to remove older snapshots. As mentioned earlier, the default threshold for transaction logs is 30 days. With VACUUM, the default is 7 days, and you can choose the retention time yourself:

```
VACUUM tzl_delta RETAIN 200 HOURS
```

Note that you shouldn't pick a too low value. You might end up in a situation where you're cleaning up files that are still being used to long-running jobs. Databricks will even warn you if you try to go below 168 hours.

Transaction logs

Databricks keeps track of all the changes you do on a Delta Lake table in a log. You can look at all the versions of the data and even look at the table for a specific point in time. So if you want to see the difference in data between two points, you can, without having to do any logic yourself. Let's create a quick test. Note that the date in the last example needs to be changed to work:

```
DESCRIBE HISTORY tzl_delta;
SELECT * FROM tzl_delta VERSION AS OF 1 MINUS SELECT * FROM tzl_delta VERSION AS OF 0;
SELECT * FROM tzl_delta TIMESTAMP AS OF '2020-01-01 10:00:00' ;
```

This is a pretty interesting feature with a lot of possibilities. You can use it to fix mistakes and errors. If someone accidentally removes a few rows, you can just look at an earlier version and use that result set to update the current one, thereby fixing the problem.

For normal operation, you might not want to keep this information around forever. The default setting is to keep it for 30 days. If you want to keep it longer, or shorter, you can. We'll talk more about that later when we go through settings.

Selecting metadata

As you've seen, SQL commands can also be used to get metadata from the system. You can get a lot of information about the objects in the database using a few simple commands, like we've already seen with DESCRIBE. This is also one of the commands you'll probably use most frequently:

```
DESCRIBE taxi_drivers
DESC detail taxi_drivers
```

The last one is yet another command to give you information about a table. It will be most useful when used in combination will Delta Lake tables. There are many ways to get at the table data as you've seen.

Another powerful command is SHOW. Just like DESCRIBE, it'll help you learn more about the objects you are working with. There are quite a few of them, but let's go through them and see what they do:

```
SHOW DATABASES
SHOW DATABASES LIKE '*taxi*'
```

This will list all the databases (or schemas, if you prefer) in the current Databricks workspace. As you can see in the second example, you can use a filter to limit the result set. Once you have the databases, you can look at the tables. Note the use of an asterisk rather than a percentage sign in the LIKE clause:

```
SHOW TABLES
SHOW TABLES FROM default LIKE '*fare*'
```

The first command will show you all tables in the current database. If you want to see the list from another one, you need to use the FROM parameter to specify which one. As with SHOW DATABASES, you can use LIKE to filter what you get back. If you want to dig deeper into the tables, you can use another command:

```
SHOW TBLPROPERTIES taxi_drivers
```

This command is only truly useful if you add table properties. Otherwise, you won't get very much information out of this. There's more information to look at if you use DESCRIBE EXTENDED instead. But it can be good to know that it exists, especially if you want to script things. You can look at the status of a given property to decide what to do next:

```
SHOW COLUMNS FROM tzl_delta
```

This will return a list of all the columns in the specified table. It won't give you the type or the comments, so unless you just want the names, you're better off using the DESCRIBE command mentioned earlier.

You can also look at the functions available to you. This will create a huge list of everything you can use in Databricks. While it is big, it's actually quite good to skim through it as you can find something useful:

```
SHOW ALL FUNCTIONS
SHOW SYSTEM FUNCTIONS LIKE '*SU*'
SHOW USER FUNCTIONS LIKE '*TAX*'
```

The SYSTEM argument will limit the list to the built-in functions. USER will only return what you built yourself; and ALL will, as you might expect, return everything. Here as well you can limit the list with the LIKE command.

As a small sidenote, I actually export this list to a separate text file to have a list of all available functions easily available at all times. There's a lot on the list I don't use frequently enough to remember from the top of my head, but going through the list I can find what I'm looking for.

Gathering statistics

There's one more thing that is good to know. When you run a query, the Databricks optimizer (named Catalyst) will look at what you want to do and create a plan of how to get the data most efficiently. Databricks will then execute the work as best it can.

Unfortunately the basic way it does this is by following a set of rules. Sometimes that's good enough. Often it isn't. Fortunately there's a cost-based optimizer that can help you. It does a better guess of how to pick up data in the fastest way possible.

To do this efficiently, the cost-based optimizer needs statistics about the table it's about to read. The more information it has about the objects, the better it can guess. You can help the optimizer by running the ANALYZE TABLE command:

```
ANALYZE TABLE yellowcab_tripdata_2019_06 COMPUTE STATISTICS;
ANALYZE TABLE yellowcab_tripdata_2019_06 COMPUTE STATISTICS FOR COLUMNS
tpep_pickup_datetime,tpep_dropoff_datetime, PULocationID,DOLocationID;
```

The first one will scan the table to gather overall information about it. The second command will go through the specified columns as well. This is a good idea to do for columns you use frequently.

Note that it is important to run the analysis at the right time. If you fill a table, analyze, and then change the dataset, you'll confuse the optimizer. Keep this in mind as bad table statistics can frequently be a cause of bad performance.

There is a way to get a preview of the plan Databricks is planning on using. This is very helpful in figuring out why a particular query takes more time than you expect. The output isn't very intuitive, but we'll come back to this in Chapter 12:

```
EXPLAIN SELECT * FROM yellowcab_tripdata_2019_06;
```

Running this command is not very useful for a simple query like this. Where it can be helpful is when you have a complex query with multiple joins. Then you want to make sure that the optimizer does its work in the correct way.

Summary

In this chapter, we got familiar with the Structured Query Language, SQL, that is being widely used for data analytics. We looked at selecting data from tables and how to filter it, order it, and combine it with other datasets.

We then used math functions on the result to figure out a bit more about the data. By using simple features in SQL, we got interesting, aggregated information about the taxi rides in the city.

Then we looked at Delta Lake SQL, the future way of working with data in Databricks. We proved that updating, deleting, and merging work perfectly fine even in a Spark world, making Databricks useful in even more scenarios.

Finally, we used some magical commands to get metadata information about our objects and the optimizer.

That was a quick introduction to SQL with Databricks. Most likely, you'll use this a lot during the exploration phase of an analytics project. But once you want to do more advanced stuff, including running algorithmic code, you'll want something else. That something else will most likely be Python. Time to start coding for real.

CHAPTER 7

The Power of Python

Python has quickly become one of the most important tools in the data science and data engineering communities. This chapter digs deeper into how you can use this language together with the Apache Spark DataFrames API to work with data in an efficient way.

We'll discuss what Python is and quickly look at how the language is structured. We will revisit DataFrames and learn how to play around with data using some of the built-in features.

With our datasets in place, we'll learn how to pick up data, filter the information we want, and run different functions to get the results we want. Then we'll look at how to read and write the data to and from file systems.

Finally, we'll look at how we can use different join statements to combine different DataFrames to create new ones or extend existing ones. We'll even look at how to create Cartesian products that can bring the biggest of servers to its knees.

Python: The language of choice

As already mentioned, you can use Databricks with a number of languages. At least if you are ready to connect via a tool. If you however want to work in the notebook user interface, there are exactly four options.

Many people use Scala. It's what Spark was written in, and for a long time it was the fastest option. Some data scientists prefer R, with its roots in statistical computing. As mentioned in the last chapter, almost everyone uses SQL to some degree. As it lacks features that come in handy when building data flows, there is another choice that has moved into standard territory.

That choice is one of the most popular alternatives in Spark – Python. It's an old language that really took off when analytics became popular. Data science–focused website KDnuggets does a yearly poll of what tools their users prefer. Python leads the charge with 65.8 percent of respondents using it in 2019, up from 59 percent in 2017. For comparison, number 2, RapidMiner, is used by 51.2 percent of their respondents.

While Python is great at data science work, it's also an excellent general language. You can use it to wrangle data, but also to write web services and computer games. So learning it gives you access to a skill you can use both when you crunch data and automate server management and when you read the pins on your Raspberry Pi.

Best of all, Python is relatively easy to learn. Code is usually readable, and it's interpreted, meaning you don't need to compile any code to run. That makes trial-and-error development quick.

There are also a huge number of packages available for you to download free of charge. Every package has some functionality that can simplify your development. All of them can be installed in Databricks and used within your notebooks.

If you go to pypi.com, the main index page, you'll find there are hundreds of thousands of projects. Use the search box to find something interesting. Pandas, scipy, and scikit-learn are three examples commonly used in data science.

All this has made Python the language you'll come across most frequently in your data journeys. This is especially true for Spark and Databricks. This is why we spend quite a few pages getting familiar with it. Let's start by getting a feel for what it looks like.

A turbo-charged intro to Python

I won't dig deep into Python basics here, but I'll blow through the most important things here so you can get going even with limited prior knowledge. If you've ever done any type of programming, you'll be just fine. Most things we'll do are not full-blown code but rather data munging.

One important thing: There are two major versions of Python available right now, 2 and 3. They have been developed in parallel for many years, but when you read this, version 2 should finally have been put to rest. Always use version 3, no matter if you code in Apache Spark or traditional software.

Like most other languages, Python uses variables, and it uses the equal sign to assign data. There are many types, like integer, double, and string. But there are also lists, tuples, and dictionaries that are useful in a data science setting:

```
i = 1
s = 'String'
l = ['This','is','a','list']
d = { 'A':5, 'B':6 }
```

Strings can be wrapped in both single quotes and double quotes. It doesn't matter which one you use. This is convenient if you want to use English language within a string, among other things. Note also that strings are arrays of characters and that you can subset them. The following second command will print out characters between positions 8 and 15, which is believe.

```
s = "I can't believe it"
print(s[8:15])
```

Branching is primarily done with if/elif/else statements. You do a main check at the top with if and additional checks with elif and capture non-hits with else. To separate the if statement from the body, you use a colon and an indent. Note that indents are required. You'll get an error if you don't use them or use them where none is expected:

```
x = 1
if x == 1:
  print('One')
elif x == 2:
  print('Two')
else:
  print('Neither one or two')
```

Loops are done with for, if you know what to iterate over, and while if you do not. Like with the if command, you need to use a colon and an indent to separate the commands from the body you want to run:

```
# This is the for loop
for x in ['Spark','Hive','Hadoop']:
  print(x)

# This is the while loop
x = 0
while x < 10:
  print(x)
  x = x + 1
```

Functions can be called in-line. You pass parameters within parentheses, arguments separated by a comma. Let's look at an example where we use a function named range to create for our loop. The first argument is the starting point, the second is the endpoint, and the third is the step:

```python
for i in range(1,10,1):
  print(i)
```

You can of course create your own functions. In its most basic form, you define the function with def, set a name, and tell it what you expect in the form of arguments. In the body, you run the code and finally you return a result, if any:

```python
def my_function(input_var):
  return(input_var * 2)

# This is calling the function.
my_function(4)
```

There are thousands upon thousands of bundled functions available in the form of packages. To use them, you need to make sure they are installed (usually with the terminal command pip) and then import them in your code. You can reference the whole package or specify exactly what you want. It's also possible to add an alias:

```python
import time
time.sleep(10)

import time as t
t.sleep(10)

from time import sleep
sleep(10)
```

Lines are flagged as comments and won't be executed if you add a # sign in front of them, as you might have noticed in the preceding code. Even though it's simple to add information about your code, few do enough of it. Please add a lot of documentation in your code. Your future self, if no one else, will thank you for it:

```python
# This won't be executed, but it also won't create an error.
def my_function(input_var):
  return(input_var*2) # This works too.
```

In Python, you'll find that functions are available pretty much everywhere. The beautiful part is that you can chain them together in a very easy way. Even seemingly trivial objects like variables have a number of modules. Strings, for instance, have tons and tons of them:

```
s = ' spring'
s.replace('p','t').strip().capitalize()
```

If you want to handle errors, there's built-in support for that as well. You can wrap your code into try/except constructs to make sure errors are caught. If you know what to expect, you can sometimes also handle them:

```
try:
  # This is where your code goes.
  1 + 1 + int('A')
except:
  # This will trigger when there's an error in the body above.
  print("You can't convert a letter to a number")
```

Yes, that was a very brief introduction. Don't say I didn't warn you. Python is a great language, and there are many thick books written about it. It's not sensible to dig deep here. Luckily, most of what we'll do in this book is Spark related, so you only need to know the basics.

Finding the data

So Python is pretty easy. Let's see if we can actually use it to play around with some data. This time we'll use files supplied by Databricks to play around with. We'll go with the US Department of Transportation's Bureau of Transportation Statistics of on-time performance for domestic flights. Now that's a name that just rolls of the tongue.

As it's stored in DBFS, we can investigate the files using the %fs commands. Let's start by looking at the folder structure. It's available under the databricks-datasets folder. Airlines is what we're looking for:

```
%fs ls /databricks-datasets/airlines
```

In the folder, you'll notice that there is a readme file. It's a pretty common thing that will show you information about the dataset. We can use the head command to look at the first rows of the content:

```
%fs head /databricks-datasets/airlines/README.md
```

Ok, this gives us a little bit of background and the source of the information. In this case, we also get all of the data with the head command. That's not always the case. If you need to see more, you can try using a shell command on the driver instead. Note that we then need to prefix it with /dbfs/:

```
%sh cat /dbfs/databricks-datasets/airlines/README.md
```

If you look further into the folder structure, you'll notice that the data files are also in the same folder. They're named part-00000 with number from 00000 to…well, a lot. Let's see if we can check the number of files with a little Python:

```python
import glob
parts = glob.glob('/dbfs/databricks-datasets/airlines/part*')
len(parts)
```

Here are a few things to unpack. If you want to look at the file system, there are many commands in Python. The walk and listdir procedures in the os package are two examples. In this case, we want to run a wildcard search, so we choose glob instead.

We run it with glob.glob. It might seem strange but just means we run the glob procedure in the glob package. The argument is the path to the file. You might notice we added /dbfs/ to the beginning of the string. This is because we're running the command on the driver. The Databricks File System is mounted to the driver under /dbfs.

The result is a long list of all the files matching part*. By using len, we can count the number of items in the list. The result is 1,920 files. That's a lot. If you have a small cluster, it'll take a long time to work with this dataset. So we'll just use a subset. Let's see how to do that. But first, let's think about where the data will go once loaded.

DataFrames: Where active data lives

While you can run Python pretty much in any way you want in Databricks, it is optimized to work with the DataFrames API. You can think of a DataFrame as a table. They have named columns with defined types, or what's commonly referred to as a schema.

CHAPTER 7 THE POWER OF PYTHON

Underneath the surface, they still build on RDDs, but unless you need something very special, stick to DataFrames.

If you're familiar with R or the Python package Pandas, it's important to remember that DataFrames are similar but not identical to dataframes in those tools. This can be slightly confusing as not all concepts translate. Things you can do in Pandas don't necessarily work in Spark DataFrames and vice versa. Syntax is also different.

The really great thing about Apache Spark DataFrames is their ease of use. As you'll notice in this chapter, you get a lot of power out of the box. You can run advanced functions on a huge amount of data with just a few commands. Apache Spark will make sure that it's running efficiently under the surface. Let's look at what a command looks like:

```
df = spark.read.parquet()
df.count()
```

This will read a Parquet file and make it accessible to you via the df variable name, representing a DataFrame. You can of course use any other name for the variable, like myDF or my_dataframe. The variable turns into a DataFrame as that's what the read function returns.

The next command is running a count on the number of rows in the DataFrame. As this is an action, the job will actually run. After a few seconds, you should see a result.

Don't forget that transformations won't trigger anything. Spark is using lazy bindings. That means it won't execute the commands until it gets a request for an action. This is important to remember when you write more complex code, as we'll see in a later chapter.

DataFrames can work both with external data and with the information you have stored in DBFS. It gladly reads from Hive and lets you use SQL to pull data back into a DataFrame.

Once the data is in there, you can do a lot of different transformations and actions to the data. Of course you can return the result to a place of your choice, something you should do before you shut down your cluster as the DataFrames only live while the nodes are spinning.

The main takeaway here is that most of the data you'll be working with in Apache Spark will, when working with it, be stored in DataFrames. So with this background, let's get started and load some data.

CHAPTER 7   THE POWER OF PYTHON

Getting some data

We know where the data is located and the format. We still don't have a proper view of the content. Just as we looked at the head of the readme file, we need to investigate the first rows in the first data file:

`%fs head /databricks-datasets/airlines/part-00000`

You'll notice that there is a header and that the delimiter is a comma. Interestingly enough, this is only true for the first file. That's at least true for the header. Anyways, let's read the first file into a DataFrame and look at it:

```
df = spark
    .read
    .option('header','True')
    .option('delimiter',',')
    .option('inferSchema','True')
    .csv('/databricks-datasets/airlines/part-00000')
```

There are a ton of options when reading data, as we've already seen in an earlier chapter. Here we start by reading the part-00000 file from the airlines folder into the df DataFrame. Options are passed to tell Spark there is a header row and a colon delimiter. Then we ask the system to infer the schema, as we don't have it in any other way. Let's make sure the schema looks fine:

`df.printSchema()`

The printSchema command will show you the DataFrame schema. If you quickly look through it, it looks just fine. As we'll see later, it's actually not that good. For now, let's leave it as is and look at the actual data instead:

`display(df)`

The display command takes the DataFrame, tidies it up, and shows it in a grid. We get the first 1,000 rows from the first file in the set. This makes it possible for us to validate the data and secure it looks as we expect it to.

If you want a smaller list of values, you can use the limit command to do so. This command can be used in many different ways and not only to show fewer rows as we

CHAPTER 7 THE POWER OF PYTHON

do in the following. It's helpful to have at hand when you just want a quick glance at the structure of a large dataset:

```
display(df.limit(5))
```

There is another command you can run to look at the data. With show, you get the result in a raw format. It won't be as nicely formatted. It can however show you things that you might miss in the cleaned-up display result. Also, display is Databricks specific. To look at an example of the general alternative, just run

```
df.show()
df.limit(5).show()
```

Now, as the dataset is pretty large, we can't just read all of the files, at least not without a large cluster. Even then it'll be a tad slow. So let's limit the data by adding the asterisk in the right place. Ten files should be fine. The following command shows how you do it, but don't execute it just yet:

```
df = spark
.read
.option('header','True')
.option('delimiter',',')
.option('inferSchema','True')
.csv('/databricks-datasets/airlines/part-0000*')
```

That's quite nifty. The asterisk doesn't even have to be in the end. You can use the wildcard character freely and create advanced patterns. This also goes for folders. So you can have something like /mydata/20*/*/daily.csv to get all data for all years between 2000 and 2099, all months, assuming that's the folder structure you're using.

The problem with our preceding command is that if the files are huge, you'll spend a lot of time inferring the same schema. That's not great. If you know the schema is the same, you can actually read just one file, pick out the schema, and then use it for the following files.

Also, as we've noticed, in this case only the first file actually has a header. So if you run the inferSchema option on all files, you'll get very odd results. This too is solved by using the technique shown in the following:

```
df = spark
.read
.option('header','True')
```

```
.option('delimiter',',')
.option('inferSchema','True')
.csv('/databricks-datasets/airlines/part-00000')

schema = df.schema

df_partial = spark
.read
.option('header','True')
.option('delimiter',',')
.schema(schema)
.csv('/databricks-datasets/airlines/part-0000*')
```

In this example, we read the first file just like in the preceding code. We then copy the schema from the DataFrame df and save it into a variable called schema. The variable is used in the next command where we instead of option use the schema command.

If you try both options, you'll notice that the latter version is faster. The only difference is the inference. What Spark actually does if you choose that option is to read through the file and look at all the fields to figure out what type it is. That takes time, especially if the files are large.

An even better alternative is of course to provide your schema, assuming you know it or want to build it. Then you don't need to go through this extra step. I'll show you how to do this, but not for the full table as it would take up more than a page:

```
from pyspark.sql.types import IntegerType, StringType, StructField, StructType

schema = StructType([
            StructField("Year",IntegerType(),True),
            StructField("Month",IntegerType(),True),
            StructField("DayofMonth",IntegerType(),True),
...
            StructField("IsDepDelayed",StringType(),True)
])
```

You import the data types from the pyspark.sql.types library. In this case we get IntegerType and StringType, but there are many more – DateType and DoubleType, for instance. You can import all of them with from pyspark.sql.types import *.

With the import done, we define the schema in the form of a StructType. The StructType expects a list of StructFields. So that's what we do next, one row at a time. In the StructField, we give the column name and the type and tell Spark if the field accepts null values. Then we have a schema we can use for import. The three dots (...) just indicate that I removed a number of rows. They are not part of the actual syntax.

Did you notice that we added empty parentheses after StringType and IntegerType? That's Python's way of indicating this is a function and not a variable or constant. If you look at StructField, it also has them. The content inside the parentheses are arguments, or parameters. We'll get back to this in a bit, when we create our own functions.

A somewhat annoying detail is that you can get the schema from any DataFrame, just as you saw earlier. But it's in Scala format, so what you get isn't directly translatable to Python. You need to change a few things to make it compatible with the structure here:

```
print(df.schema)
```

For us right now, the schema we got from the inference is good enough. We have the data in our DataFrame, and it's time to start playing around with it. Let's start by just checking what it looks like and how many rows it contains. This will give us a feeling for the data:

```
display(df.dtypes)
df.count()
```

Did you notice that the count took a little bit of time? That's the lazy evaluation working. It's not until you hit an action that stuff starts happening. The first command just looks in the Hive Metastore for information and doesn't have to read the data. Pretty smart. Run the display command again to look at the data:

```
display(df)
```

Short note here: Feel free to read even fewer files if you work with a small cluster and don't want to wait. While this dataset isn't humongous, with more than 1 billion rows, it's still large enough to make Spark break a sweat. The commands we'll use will work just fine even if you just pick up data-00000.

Selecting data from DataFrames

So, with that out of the way, let's start by only looking at year, month, dayofmonth, arrdelay, and depdelay. This is filtering on the column level and usually a good thing. Only read the data you need. You can do that in a number of ways, but let's try a couple:

display(df['year', 'month', 'dayofmonth', 'arrdelay','depdelay'])

display(df.select(['year', 'month', 'dayofmonth', 'arrdelay','depdelay']))

You can use either method. It really makes no difference. In some cases it's easier to use one instead of another from a syntax perspective, but it's mostly a style choice. I personally prefer to go between the different alternatives in different situations though.

No matter what syntax you choose, plain select isn't very interesting. Let's do some aggregation. It would be interesting to know what the average delay time was across the different months:

display(df.groupBy('Month').avg('arrdelay'))

That didn't work! That's strange, right? Luckily the error text points us in the right direction: ArrDelay is not a numeric column. Let's go back and look at the schema. It turns out that our inferring didn't go so well, after all. The arrdelay column is actually parsed as a string, not as a number. That probably means we have some non-numeric values in there.

We'll get to how you can find unexpected rows like that with code in a later chapter, but for now just scroll through the 1,000 rows that the display command gives you. After a little bit, you'll find the culprit. It seems like some pesky NAs have found their way into the data.

So if we want to run calculations on that field, we need to clean those up. There are many ways to do that, and we'll talk more about that in Chapter 8. Right now we'll just do a dirty convert:

```
from pyspark.sql.types import DoubleType
from pyspark.sql.types import DoubleType

df = df
    .withColumn("ArrDelayDouble",df["ArrDelay"]
    .cast(DoubleType()))
```

We again import the DoubleType as that's the target format we want. Then it's time to start processing. The withColumn command is used to either add a new column or replace an existing one. While we could just replace the ArrDelay, it might be good to keep the original around in case we need to look at it later.

So in this case we create a new column named ArrDelayDouble. The second argument decides what goes into the column. We use the ArrDelay column as a source, but convert – or cast – it to a new data type. The result is sent right back to df.

If we look at the columns now, you'll see that NA has turned into nulls. And that's accepted in columns of the double type. Now let's see if we can rerun the query we ran before, hopefully without any errors:

display(df.groupBy('Month').avg('arrdelaydouble'))

It works. And we get a nice average from year to year. Now, there is one interesting detail here to note. We had null values in the data. So what average did we actually get? What percentage of the rows have nulls?

WORKING WITH NULLS

Nulls can confuse. I've already mentioned it but here I go again. They can really mess things up. In the example where we cleaned up NAs, we actually didn't look at how frequent they were. That means that the averages aren't as reliable as we might think:

```
%sql
create table if not exists null_test (
        a integer,
        b integer);
insert into null_test values (1,2);
insert into null_test values (1,null);
insert into null_test values (1,2);
select avg(b) from null_test;
```

What do you expect? The answer is 2. Now, there are three rows in there, so it's not crazy to think that you would get (2 + 0 + 2)/3 = 1.33. But that's not what happens. Null isn't zero. It isn't anything. If you're not careful, this can throw your numbers off. Keep track of your nulls and handle them before doing math.

Chaining combo commands

You might have noticed something in our initial dive into Python and DataFrames. The commands kind of stack on top of each other. This makes it easy to build chains of things you want to happen. It's also pretty flexible:

```
display(df
        .select(['Year','ArrDelayDouble'])
        .groupBy('Year')
        .avg('ArrDelayDouble'))

display(df['Year','ArrDelayDouble']
        .groupBy('Year')
        .avg('ArrDelayDouble'))

display(df
        .groupBy('Month')
        .avg('arrdelaydouble'))
```

You can keep building chains like this in a rather flexible way. Position matters though. Python won't automatically figure out in which order you should run your stuff. So beware.

Also note that just like select before, all these variants give the same result. It's a matter of which syntax you prefer. Try to go for clarity whenever you can. In this case, the select part adds nothing in my opinion.

Let's look at our groupBy function and see if we can extend it. We've already seen we can use the mathematical function avg. There are a lot more to choose from. We have min, max, floor, ceil, cos, sin, and many hundreds more available through a huge number of modules.

We'll spend a lot of time looking at the pyspark modules we've already had a chance to work with, especially the SQL parts with which we can build queries in Python similar to SQL. To access them, we need to import them:

```
from pyspark.sql.functions import *
```

With the asterisk, we tell Python that we want everything that's available. If you prefer to be explicit, you can instead just ask for the functions you plan on using. For instance, you can write like this to import only the functions for col and year:

```
from pyspark.sql.functions import col, year
```

Working with these functions using DataFrames is in many ways similar to how you work with SQL. If you look at the code long enough, you'll be able to quickly translate between the two environments in your head. Let's look at a couple of examples of how to run some chains:

```
display(df
        .filter(df.Origin == 'SAN')
        .groupBy('DayOfWeek')
        .avg('arrdelaydouble'))
```

Filtering is one of the most common operations you'll do. In this query we're asking for rows where the origin is San Diego Airport, airport code SAN. As you see we do this before we group the data. We also use the direct mentioning of the column. If you prefer, you can use the bracket format instead:

```
display(df.filter(df['Origin'] == 'SAN')
        .groupBy('DayOfWeek')
        .avg('arrdelaydouble'))
```

You can use multiple filters if you want. Let's say you're only interested in planes leaving from San Diego to San Francisco. Then you need to filter on both the origin and dest columns. That looks like this. Note that you need the parentheses:

```
display(df
        .filter((df.Origin == 'SAN') & (df.Dest == 'SFO'))
        .groupBy('DayOfWeek')
        .avg('arrdelaydouble'))
```

The & character tells us both statements need to be true: X and Y. There are other operands as well. With |, either statement or both statements can be true: X or Y. It's of course also possible to use more than just equal signs in the filters.

```
display(df
.filter(((df.Origin != 'SAN') & (df.DayOfWeek < 3)) | (df.Origin == 'SFO'))
.groupBy('DayOfWeek')
.avg('arrdelaydouble'))
```

In this somewhat hard to read filter, we are looking for planes that are leaving from any airport other than San Diego in the first two days of the week plus all planes leaving from San Francisco. As filtering gets more complex, it's usually better to break them down into multiple commands instead:

```
display(df
        .filter(df.Origin != 'SAN')
        .filter(df.DayOfWeek < 3)
        .groupBy('DayOfWeek')
        .avg('arrdelaydouble'))
```

Splitting logic up like this usually makes it easier to read. It also lowers the risk of making mistakes. Anyways, you might have noticed that the results come back a bit unordered. Let's see if we can sort them by day of the week:

```
display(df
        .filter(df.Origin == 'SAN')
        .groupBy('DayOfWeek')
        .avg('arrdelaydouble')
        .sort('DayOfWeek'))
```

This will sort the result set based on the DayOrWeek column. We can see there is a little dip on Saturdays, which is nice to know when booking a flight (in 1987.) Like always in Python, there are other ways to sort. You can use orderBy instead in this case, with the same result:

```
display(df.filter(df.Origin == 'SAN')
        .groupBy('DayOfWeek')
        .avg('arrdelaydouble')
        .orderBy('DayOfWeek')
```

Whatever method you prefer, the result is still a bit hard to read with all those numbers. Let's round them to only one decimal point. To do that easily, we need to use a new way of running averages. We also need to import a few functions. Let's have a look:

```
from pyspark.sql.functions import mean, round

display(df.filter(df.Origin == 'SAN')
        .groupBy('DayOfWeek')
        .agg(round(mean('arrdelaydouble'),1)))
```

Agg gives us an ability to run functions on the aggregate we want to use as it returns a DataFrame. In this case we run an average using the SQL function mean and then pass it along to a rounding function. Round takes a second argument that defines the number of decimal points. In this case we picked one.

While the result is a little bit better, the naming of the aggregate column isn't very good. Let's change that by using the alias command. That will make it much easier to read. Another plus is that you can refer to the column by the name. Let's change the name first:

```
display(df.filter(df.Origin == 'SAN')
       .groupBy('DayOfWeek')
       .agg(round(mean('arrdelaydouble'),2)
       .alias('AvgArrDelay')))
```

Now that we have the alias, let's try sorting the data. This time we want to see what days are the worst. So we need to sort in a descending order. To do this we need to import yet another function:

```
display(df
       .filter(df.Origin == 'SAN')
       .groupBy('DayOfWeek')
       .agg(round(mean('arrdelaydouble'),2).alias('AvgArrDelay'))
       .sort(desc('AvgArrDelay')))
```

There we go. It looks like it's better to travel on Saturdays instead of Thursdays. At least if you want to arrive on time in this historical time period. Let's look at the spread as well. The agg function comes to our rescue again. We only need to import another couple of SQL functions to do our math:

```
from pyspark.sql.functions import min, max

display(df
       .filter(df.Origin == 'SAN')
       .groupBy('DayOfWeek')
       .agg(min('arrdelaydouble').alias('MinDelay')
          , max('arrdelaydouble').alias('MaxDelay')
          , (max('arrdelaydouble')-min('arrdelaydouble')).alias('Spread'))
)
```

CHAPTER 7 THE POWER OF PYTHON

As you can see, the agg function can be quite nifty to have at hand. You'll probably use it a lot in your data exploration sessions. Let's continue by improving our table a bit. Year, month, and dayofmonth are nice, but a date field would help for some functions. So let's create one:

```
from pyspark.sql.functions import concat, to_date
df = df
     .withColumn('DayDate', to_date(concat('Year','Month','DayOfMonth'),
     'yyyyMMdd'))
```

Voilá! A new, proper date column. Concat combines values. In this case we create one string out of the year, month, and dayofmonth columns. We then use that string as input to the to_date command. The yyyyMMdd defines the format of the string we created.

DATE FORMAT PATTERNS

There are many ways to define a date, and it's not always obvious how to do it. It's easy to slip and type m instead of M, getting you the minute instead of the month. Here is a list of the pattern syntax.

Letter	Meaning	Example
y	Year	2020
G	Era	AD
M	Month	01
d	Day in month	12
h	Hour (1–12)	14
H	Hour (0–23)	12
m	Minute	13
s	Second	44
S	Millisecond	123
E	Day in week	Monday

(continued)

Letter	Meaning	Example
D	Day in year	101
F	Day of week in month	2
w	Week in year	33
W	Week in month	3
a	Am/pm flag	PM
k	Hour in day (1–24)	22
K	Hour in day (0–11)	10
z	Time zone	Central European Standard Time

As an example, you can reference a date like 2020-01-01 as yyyy-MM-dd. If you work with time, you might use something like HH:mm:ss:SSS for 14:02:30:222. There's a lot you can do with dates.

Feel free to look at the table with the commands we've already gone through, like dtypes, select, and limit. If you do, you'll notice that you have a properly formatted date in the column. Let's see if we can run a couple of functions on it. First, let's get the weekday in plain text:

```
display(df
    .select(date_format('DayDate', 'E').alias('WeekDay'),
    'arrdelaydouble', 'origin', 'DayOfWeek')
    .filter(df.Origin == 'SAN')
    .groupBy('WeekDay','DayOfWeek')
    .agg(round(mean('arrdelaydouble'),1).alias('AvgArrDelay'))
    .sort('DayOfWeek'))
```

This is more like it. In this example we again run the date_format function to get the date structure we want. This time we pick up the weekday name. Unlike DayDate, we don't materialize the result into the table using withColumn. Instead we do it on the fly, every time we run the query.

Let's plot this. Just below the result, on the lower-left side, click the bar chart button. Then click Plot Options. Make sure you have WeekDay in the Keys box and AvgArrDelay in the Values box. Click Apply and look at the result. Not too bad. Let's do another query and plot a line chart across a given month. If you want to see the time span you have in your DataFrame, you probably know how to do that now. If you want a range however, you need yet another function:

```
from pyspark.sql.functions import date_add

start_date, end_date = df
      .select(min("DayDate"),date_add(min("DayDate"),30))
      .first()
print(start_date)
print(end_date)
```

This time we use date_add to get a 30-day window. We're looking for the first day in the dataset with the min function. Then we do it again, but add 30 to the result with date_add. This gives us a result about a month in the future.

Note by the way that we use print to show the result and not display. The latter expects a DataFrame and can't handle other variables. You'll get an error like "Cannot call display" if you try it.

Now, let's use the variables we just created in the next query. What we want is to pick up rows based on the date range we created earlier. We can do that with a built-in command, of course:

```
display(df
      .filter(df.Origin == 'OAK')
      .filter(df
            .DayDate.between(start_date,end_date)
            )
      .groupBy('DayDate')
      .agg(mean('ArrDelay'))
      .orderBy('DayDate'))
```

The new thing here is the between command. We specify two dates in the filter. You can just type them out if you want to, like "1987-10-01", but in this case we use the variables we created in the preceding code instead.

Do a line graph on the data to see how things look during the month. Make sure that DayDate is in the Keys box and that ArrDelay is in the Values box. Drag your mouse pointer over the line to get the actual values for each point. The built-in graph engine isn't great, but it's surprisingly good considering how easy it is to use:

```
display(df
        .filter(df.Origin.isin(['SFO','SAN','OAK']))
        .filter(df
              .DayDate.between(start_date,end_date)
              )
        .groupBy('Origin','DayDate')
        .agg(mean('ArrDelay'))
        .orderBy('DayDate'))
```

Run this and put the Origin column into the Series groupings box in Plot Options. This will give you a comparison of SFO, SAN, and OAK. Notice the use of isin and the list. You can create that list as a variable if you wish, like this, for instance:

```
airport_list = ['SFO','SAN','OAK']
```

You can also create lists from DataFrames if you want to. Let's pick up five random airports from our DataFrame. We use the collect function to get the actual values instead of creating a new DataFrame:

```
airport_list = [row.Origin for row in df.select('Origin').distinct().limit(5).collect()]
```

There's a lot going on here. First of all I'm using list comprehension. We've talked about this earlier in the book. It's basically a compact for loop. We pull a distinct list of all values in the Origin column. Then we pick the top five with limit.

Now we get to the collect part. It'll return each row as a…well, row. As we want a clean list, we pick out the actual value and put it into a list. Now we can use this variable in the example we ran in the preceding code:

```
display(df
        .filter(df.Origin.isin(airport_list))
        .filter(df
              .DayDate.between(start_date,end_date)
              )
        .groupBy('Origin','DayDate')
        .agg(mean('ArrDelay'))
        .orderBy('DayDate'))
```

Feel free to go through the steps manually to figure out why we did the loop. It can be a bit confusing before you get the hang of it. Run the following commands to see what's happening. We start with a DataFrame, use collect to get a list, and need to pull the values out of the list to get a new list with clean values:

```
print(type(df.select('Origin').distinct()))
print(type(df.select('Origin').distinct().limit(5).collect()))
print(df.select('Origin').distinct().limit(5).collect())
print(df.select('Origin').distinct().limit(1).collect()[0].Origin)
```

If you find the preceding list comprehension confusing, it's possible to rewrite the code as a normal loop. It's easier to read but takes up a few more lines of code. I usually recommend using the classical loops for code you're running in production. More people are used to them and can immediately understand what's happening:

```
airport_list = []
for row in df.select('Origin').distinct().limit(5).collect():
    airport_list.append(row.Origin)
```

Let's add another column so that we can compare states. Now, we should actually do this for all airports to do this properly, but let's not as that would require a lot of text with no added educational value. So let's just pick three busy airports per state: LAX, SFO, and SAN for California; JFK, LGA, and BUF for New York; and DFW, IAH, and DAL for Texas. The rest will be flagged as OTHER. Let's see how we can do that:

```
from pyspark.sql.functions import when
df = df.withColumn('State',
                   when(col('Origin') == 'SAN', 'California')
                   .when(df.Origin == 'LAX', 'California')
                   .when(df.Origin == 'SAN', 'California')
                   .when((df.Origin == 'JFK') | (df.Origin == 'LGA') |
                   (df.Origin == 'BUF'), 'New York')
                   .otherwise('Other')
                   )
```

As you can see, we use the same logic as before (albeit two different versions). The main difference is that we use when and otherwise to control what should go into the new column, State. You can chain however many when statements as you wish. It's also possible to wrap multiple checks in a single when call. The otherwise function will return the default value for all non-matching rows. Let's have a look at the result.

```
display(dfgroupBy('State').count())
```

You might have noticed that we didn't do Texas. My mistake. Let's try to fix that with the same construct. Of course we don't want yet another column, so we have to update the DataFrame while keeping the old values intact:

```
from pyspark.sql.functions import col

df = dfwithColumn('State',when(col('Origin') == 'DFW', 'Texas').
when(col('Origin') == 'IAH', 'Texas').when(col('Origin') == 'DAL',
'Texas').otherwise(col('State')))
display(df2.groupBy('State').count())
```

Note that we're using col instead of the direct reference in this example. It's a function that returns the content of the given column. So here it does the same this we did previously, just using less characters.

Now, if you want to just add a column with the same value for all rows, you can't just add it with plain text. If you try something like df.withColumn('NewCol', 'Myval'), it'll fail. While you could create a when/otherwise construct, there's a better way:

```
from pyspark.sql.functions import lit

df = df.withColumn('Flag', lit('Original'))
```

The lit function returns a column with a single value. This can be helpful in some situations, as we'll see. For now you should just keep in mind that when we do operations like these, the expected input is columns, not single variables.

Working with multiple DataFrames

So far we've updated the same DataFrame over and over. It's of course possible to create new DataFrames based on existing ones. Let's create one containing a subset of the original one, filtered on just one airport:

```
df_dfw = df
        .select('DayDate','Origin','Dest')
        .filter(col('State') == 'California')

display(df_dfw.groupBy('Origin','Dest').count())
```

This is a small DataFrame you can use if you're interested in how many flights left California airports for other airports across the country. Looking at those values over time might actually give you some insight into how interest for areas comes and goes over time.

It would be nice if we could create buckets, or bins, to segment the data into chunks like small, medium, and large. For fun, let's create arbitrary limits at 400 and 1,000 per month. In reality it would probably be wise to look at the data, but let's keep it simple.

There are many ways to solve this. One way is to create a user-defined function, or UDF for short. It's a function you write yourself. This gives you the opportunity to create any type of logic really. Let's look at an example:

```python
def bins(flights):
  if flights < 400:
    return 'Small'
  elif flights >= 1000:
    return 'Large'
  else:
    return 'Medium'

from pyspark.sql.functions import count, mean

df_temp = df
  .groupBy('Origin','Dest')
  .agg(
    count('Origin').alias('count'),
    mean('arrdelaydouble').alias('AvgDelay'))

from pyspark.sql.types import StringType

bins_udf = udf(bins, StringType())
df_temp = df_temp.withColumn("Size", bins_udf("count"))
```

Before I explain what is going on, I have to tell you this is not the way to do this. There are multiple ways to solve this, and an UDF is probably the worst. That's true in general, by the way. If there is any other way to solve what you want to do, you shouldn't create your own function. They are performance killers of the worst kind. That said, they do have their place, just not as often as people think. As for this code, I'll show you a preferred way in the advanced section of the book.

Ok, with that said, let's get to it. First, we create a function called bins. It accepts one parameter, flights. It looks at the number of flights and returns a string indicating the size of the airport.

Next up, we create our play DataFrame with the origin, dest, count, and arrival delay mean. Note that we name the count column. We then define our function as an udf, telling Databricks it'll return a StringType. With this ready, we can create a new column in our temporary DataFrame.

As you can see, we can use it like any other function, only we have to go via the intermediary bins_udf. We send the count variable, it gets processed in bins, and back comes a string that we put into the new Size column.

We can then use this new column to look how delays are different between airport sizes. While this is a somewhat crude example, binning is often an interesting way to look at data and how different parts of it differ:

```
display(df_temp.groupBy('Size').agg(mean('AvgDelay')))
```

By the way, you might think that the bins function is too expressive. It's easy to make it more compact, but remember that you frequently throw readability out the window when you try to save lines. This code does the same work, but is in my opinion much harder to read. I add it here mostly so you can see the iif structure available in Python:

```
def bins(flights):
  ret = 'Small' if flights < 400 else ('Large' if flights >= 1000 else 'Medium')
  return ret
```

Another thing you might come across is situations where you need to create a DataFrame from scratch. This might happen when you get manual data that you type in. While it's not great, it can happen, especially for very small lookup tables:

```
dfg = spark.createDataFrame([
  ['AA','American Airlines'],
  ['DL', 'Delta Airlines'],
  ['UA', 'United Airlines']
], ['Shortname','Fullname'])

display(dfg)
```

With createDataFrame, we can quickly create a small DataFrame. You basically just type it out in the form of a list of lists. Not very convenient, but it's sometimes handy. So at least keep in mind that you have the option.

Another thing you might want to do is to make the DataFrames available for SQL use. There are a number of ways to do that. On the highest level, you need to decide if you want them to persist across cluster startups or not. In many cases, you might be happy with just having a temporary view while you work with it:

```
df.createOrReplaceTempView('AirportData')
df_test = spark.sql('select * from AirportData')

%sql
select count(*) from AirportData;
```

We start by creating a temporary view. The OrReplace part will make this command rerunnable without an explicit drop. If there already is a view with the same name, it'll be overwritten. If you don't want this, just try createTempView, which will fail if the view is already available.

You can of course get rid of it as well. It'll die with the cluster, but if you want it gone sooner, you can run the dropTempView command. It just takes the name of the view as a parameter.

Next up in the preceding example, we run a SQL query on the view two different ways. As you can see, this is a good way to go between Python and SQL. While most people tend to stick with the languages they are familiar with, it's frequently useful to move between them. Some things just work better in a given language.

Note that views are just a virtual pointer to the actual query. If you create a complicated DataFrame with lots of processing, every call to the view will redo the work. It's just like if you sent an action to the DataFrame directly.

If you want to share the view across sessions, you need to use the createOrReplaceGlobalTempView and createGlobalTempView instead. Note that they are still temporary and will be destroyed when the cluster shuts down.

To persist the data in Databricks, you need to create tables. Luckily that is easy and very similar to the preceding commands. The core command is write, and it gives you a lot of options. Let's start by looking at how we can create a table:

CHAPTER 7   THE POWER OF PYTHON

```
spark.sql('create database airlines')
df.write.saveAsTable('airlines.subset')

%sql
select count(*) from airlines.subset;
```

We start by creating a database. Unless we store our tables in a named database, they'll end up in default. While that works, it's easy to just fill the default database with crap. Next up, we use saveAsTable and specify both database and table names with the dot notation. Once it's stored, we can select from it like any other table. It's also persistent, of course.

One important difference is that the once the result set is stored, you disconnect it from the DataFrame. The information is stored into a table, and the next time you access it, you'll get it from the table and not the original source.

Imagine that you read from all the CSV files in the airlines dataset and create a filter that only looks at a very small subset, like 100 rows. If you use the original DataFrame or a view, you'll do the processing from scratch. With the table, you'll only look at the 100 rows in the original result set. You won't, however, get any new data that might have been added. If you want to try it and see the difference, run these commands:

```
df_bigger = spark
        .read
        .option('header','True')
        .option('delimiter',',')
        .schema(schema)
        .csv('/databricks-datasets/airlines/part-000*')

        df_sck = df_bigger.filter(col('Origin') == 'SCK')
        df_sck.createOrReplaceTempView('SCK_V')
        df_sck.write.saveAsTable('airlines.sck_t')

%sql select count(*) from SCK_V;

select count(*) from airlines.SCK_T;
```

Did you notice the time difference? Try the last two statements a few times to get a good average. Even with this small dataset, the persisted table was almost 20 times faster than reading the view, which makes sense. Picking up a few thousand rows is easier than processing millions of them, even for Spark. This is, by the way, a common trap. I'll come back to the topic when we talk about putting code into production.

CHAPTER 7   THE POWER OF PYTHON

You might also want to store the data externally. Not everything needs to be stored in Databricks. There are a number of ways to write to the file system. Let's look at how we can store data in a few common formats. You'll notice there are a few caveats, so let's look at CSV first:

```
%fs mkdirs /tmp/airlines

df.write.csv('/tmp/airlines/alcsv', sep='|', header='True')

%fs ls /tmp/airlines/alcsv/
```

It starts out just normal. Create an airlines folder and use the write function to send a CSV file to the folder. We use our own delimiter, or separator, and also tell the function that we want headers. Then things start to get weird.

The data wasn't written to the single file that you expected, but rather you get a number of files into a directory. This is a somewhat strange behavior that has to do with how Spark stores the data. As you might recall, the information is partitioned. Let's take a closer look at our dataset:

```
df.rdd.getNumPartitions()
```

As you might recall, DataFrames are nothing but a front. Underneath it's still our good old RDDs at work. We can reference them and ask how many parts our data is sliced up in. What you'll notice is that it matches up with the number of files in the folder we wrote to earlier.

Having the data in multiple files might actually be a good idea, depending on your needs. The good news is that you can control it by repartitioning the DataFrame according to your needs:

```
dbutils.fs.rm("/tmp/airlines/alcsv", recurse=True)

df
    .repartition(1)
    .write.csv('/tmp/airlines/alcsv', sep='|', header='True')

%fs ls /tmp/airlines/alcsv/
```

As mentioned before the %fs commands can be called using the dbutils package. So we do that here as to easily tell it to remove our folder. The recurse flag tells Databricks we want to remove the folder and the files in it recursively.

Then we run the same command as before, with one difference. We've added the repartitioning command in there. With that we shrink the number of partitions from four to one before we write. Note that this is only for the write. The df DataFrame will stay at the original number of partitions.

As you'll see in the folder, you'll still have a number of control files in there. There will only be one CSV file though. And that's where your data resides. The name will be garbled, but you can of course rename it if you want:

```
from glob import glob
filename = glob('/dbfs/tmp/airlines/alcsv/*.csv')[0].replace('/dbfs','')

dbutils.fs.mv(filename, '/tmp/airlines/alcsv/mydata.csv')
```

Like before, we use glob to pick up the file we want. The [0] indicates we want the first item in the list (and there should only be one CSV file), and replace removes the /dbfs part. As you might remember, we need to handle commands on the driver node differently from how we run stuff internally in Databricks. The final dbutils command does the move for us.

Do not forget that repartitioning is a costly operation. We also put all the load to a single worker which is rarely a good idea when the data increases in size. It'll be slow. In many cases it's better to leave them as is or use another tool to append the files. Spark was made for running in parallel, not in a single thread.

Now, with all this behind us, let's try another few export formats. This time we can look at a couple of the more commonly used binary formats. As mentioned before, Parquet and ORC are the ones you'll come across most frequently:

```
df.write.parquet('/tmp/airlines/alparquet/')
df.write.orc('/tmp/airlines/alorc/')
%fs ls /tmp/airlines/alparquet/
%fs ls /tmp/airlines/alorc/
```

As you can see, there is nothing much different from writing to CSV. There are functions ready to handle the data export. When we get to Avro, things change however. Let's look at how that works:

```
df.write.format("avro").save('/tmp/airlines/alavro/')
%fs ls /tmp/airlines/alavro/
```

CHAPTER 7   THE POWER OF PYTHON

In this case you need to use the format command to define that the output should be Avro. While the command is different, the result looks pretty much the same as the rest of them. Now there's just one more file type we should test. JSON is pretty much identical to CSV:

```
df.coalesce(1).write.json('/tmp/airlines/aljson')
%fs ls /tmp/airlines/aljson/
```

I use coalesce in this code. The result in this case is identical to repartition although it works differently. We'll get to that later in the book. For now, the important part is that you have files available so that we can reverse the operation and read them.

All the files we've saved so far can be read by using simple commands. CSV we've looked at. For binary files, you usually don't even have to work with parameters. Just read them. Avro is different here as well:

```
df = spark.read.parquet("/tmp/airlines/alparquet/")
df = spark.read.orc("/tmp/airlines/alorc/")
df = spark.read.json("/tmp/airlines/aljson/")
df = spark.read.format("com.databricks.spark.avro").load("/tmp/airlines/alavro/")
```

Note how we can reference the folder instead of specific files. Databricks understands what we want and picks up the data inside the folders. When you have manually partitioned files, you might want to pick up a subset, as we did in the beginning of this chapter, but this is convenient for reading everything.

Slamming data together

So far we've been working with a consolidated dataset. Many times data isn't as neatly consolidated for you. You'll need to do the merging and joining yourself. While this is one thing that SQL is really good at, you can of course do it in Python as well.

There are three basic operations we can do. First, we can add more rows to the original dataset. Second, we can add more columns or attributes to the original. Third, we can combine two different datasets that have things in common.

Let's pick up a couple of files from the original dataset and see how we add them to a common DataFrame. What's important to remember here is that both files are identical in structure. That's necessary for the union command to work:

```
df = spark
.read
.option('header','True')
.option('delimiter',',')
.option('inferSchema','True')
.csv('/databricks-datasets/airlines/part-00000')

df2 = spark
.read
.option('header','True')
.option('delimiter',',')
.option('inferSchema','True')
.csv('/databricks-datasets/airlines/part-00001')

df_merged = df1.union(df2)
```

In this example we read two different files into df1 and df2, respectively. We then combine them by taking df1 and running union with df2 as a parameter. This will take all the rows from df2 and add them to the ones in df1 – with duplicates, in case there are any.

SQL aficionados should beware of the unionAll command that also exists. In SQL there's a difference between union and union all. In Pyspark there is not. Neither removes duplicates, and the unionAll command is considered to be deprecated, unfortunately, as the result is what you'd expect from a union all in SQL.

Now assume we'd like to remove the df2 rows from the df_merged DataFrame. This is not an all too common operation, but it happens from time to time. We can do that using the exceptAll command:

```
df_minus = df_merged.exceptAll(df2)
```

A more common operation is to add additional columns to an existing DataFrame. You frequently generate or get extra data to enrich the information you already have. We've already seen this many times in this chapter, but let's create another one:

```
df1 = df.withColumn('MilesPerMinute', col("Distance") /
col("ActualElapsedTime"))
```

If you can create data based on existing data, this is an efficient method. Frequently however you need to add external information. That creates a larger problem as you need to explain to Spark how the DataFrames are connected. You do that with something called a join.

Let's create a couple of miniscule datasets so you can actually look at what happens. While joins aren't hard, it can be slightly confusing if you are not familiar with relational databases. Also, running joins on large datasets can take a bit of time, especially when you are running on very small clusters:

```
df_airlines = spark.createDataFrame([
  ['AA','American Airlines'],
  ['DL', 'Delta Airlines'],
  ['UA', 'United Airlines'],
  ['WN', 'Southwest Airlines']
], ['Shortname','Fullname'])

df_hq = spark.createDataFrame([
  ['AA','Fort Worth', 'Texas'],
  ['DL', 'Chicago', 'Illinois'],
  ['UA', 'Atlanta', 'Georgia'],
  ['FR', 'Swords', 'Ireland'],
], ['Shortname','City', 'State'])

df_cities = spark.createDataFrame([
  ['San Fransisco'],
  ['Miami'],
  ['Minneapolis']
], ['City'])
```

Great, now we have three DataFrames with a few rows each: one with the airline names, one with some additional attributes, and one with some random cities. Let's start by combining the first two in a way so we can get all full airline names to the DataFrame:

```
df_result = df_airlines.join(df_hq, df_airlines.Shortname == df_hq.Shortname)
```

We use the join command to connect the airlines data with the HQ data. The second argument tells Pyspark on which column we need to match. That means that Spark will look at a row, find a match in the other table, and know they belong together.

The problem with this statement is that we'll get the Shortname column twice in the resulting DataFrame. That's not too great as both you and Spark will get confused as to which column is which. Luckily there are a number of solutions, two of which we'll look at:

```
df_result = df_airlines
.join(df_hq.withColumnRenamed('Shortname', 'SN'), col("Shortname") == col("SN"))
```

```
df_result = df_airlines.join(df_hq, ['Shortname'])
```

Both of these will combine the data for us. The difference is that the first solution will keep both columns, just with different names. The withColumnRenamed function changes the name on the column, and we can then keep it around.

The second solution uses a special feature that automatically maps identically named columns. This was added as this problem is something you'll bump into a lot. You might notice that we put the column name in a list. If you need to map multiple columns, you can add them all into the list.

By the way, did you notice that we didn't see Southwest Airlines in the result set? Nor the Irish company Ryanair? That's because we did what is called an inner join. That means that only matching rows are shown. As we don't have corresponding data for Southwest Airlines in the df_hq DataFrame and any information for Ryanair in df_airlines, we get no result. There's a way to handle that when we want to:

```
df_result = df_airlines.join(df_hq, ['Shortname'], 'left')
```

We add the value left as the third argument, the default being inner. This shows all the rows from df_airlines and only matching rows from the df_hq DataFrame. Say hello to Southwestern Airlines. Notice that the rows for which there is no data are set to null. This is a very common way null values are introduced into datasets. Let's try joining the other way around:

```
df_result = df_airlines.join(df_hq, ['Shortname'], 'right')
```

Now we get Ryanair's headquarters information, but not all rows from the airlines DataFrame. If you want all rows from both datasets, there's another keyword you can use. Let's try that as well:

```
df_result = df_airlines.join(df_hq, ['Shortname'], 'outer')
```

All our data is finally available. You'll need to use different joins at different times. Just beware of those pesky nulls. They can cause you a lot of pain if you're not careful. It's easy to overlook and get bad data.

Now, there is another join worth mentioning. It's the dark magic of joining data, only to be used with utmost care. It's the Cartesian product, or the cross-join. It matches all the rows in one DataFrame with all the rows in another DataFrame.

Usually you don't want to use this, and when it's done in SQL, it's frequently done by mistake. Here, in Python, it requires a totally different function call. Let's see if we can use it for something clever:

```
df_result = df_cities.crossJoin(df_cities.withColumnRenamed('City', 'Dest'))
```

This would be a simple way to find all the possible routes between the cities. Not bad, but notice how joining two DataFrames with three rows turned into nine rows. With four rows, we'd have gotten 16 rows. With five we'd get 25 and so on. Now imagine if you ran this on a DataFrame with millions of rows... It's a great way to crush the poor nodes in your cluster while getting a useless result.

By the way, you probably don't want to fly to and from the same city. You could easily filter that out. I think you know how by now, but I'll show it as a small recap of where we started. For fun I'll add another syntax as well:

```
display(df_result.where(col('City') != col('Dest')))
```

Yes, you can use where instead of filter. With this you get a list of all possible combinations that aren't an identical match. Let's leave at that reminder of how easy it is to do quite powerful things in Python. Next up, we'll look at even more ways of working with data.

Summary

Whoa... That was a long one, but we're finally through it. Hopefully you'll have a better intuition of Python in general and its use combined with DataFrames in particular. That's the core combo in the Apache Spark world.

We started this chapter with a brief introduction to both Python and DataFrames. After that we dove into the data sea and pulled out some airlines data that we played around with using a large number of functions. We both sliced and diced the data with filters and summations.

The power of combinations was the next topic. Python makes it possible to build long chains of functions on top of each other. Step by step, we built up the logic we were looking for.

Finally, we looked at how to combine different DataFrames using joining and merging. This is something we'll see more of in the upcoming chapter, which will dig deeper into data crunching.

In the next chapter, we'll dig deeper into Python when we revisit ETL (Extract, Transform, Load).

CHAPTER 8

ETL and Advanced Data Wrangling

In this chapter, it's time to dig a little deeper into Python tricks that'll make your life easier. We'll revisit a lot of topics that we've already talked about, but take them a step further. First up, we'll remind ourselves of why this is important.

After we've reacquainted ourselves with ETL, we'll look into the Spark UI and how that tool can help us monitor what's happening in the system when we run a query. Then we'll take a deep dive into a lot of new functions and features available in Pyspark.

Finally, we'll look at how to handle data stored on the file system. In this part, we'll talk about managed and unmanaged tables, save modes, and partitioning. With this under our belt, we're ready to handle most data engineering tasks.

ETL: A recap

While data science and analytics is what's getting most of the attention, it's actually in the data cleansing parts most time is spent. In many projects, up to 90 percent of the time is spent understanding and restructuring data. You might hear it referenced as data engineering or data wrangling, but the work has traditionally been called ETL-ing.

We already talked about ETL, Extract/Transform/Load, earlier in this book; but let's do a short recapitulation just in case you forgot about it. Note that you might find references to ELT from time to time. It's the same letters, just in a different order to denote a slightly different process and architecture.

Extraction is the process of getting data from a source system. You might get files on a schedule or pull data on demand using something like JDBC. The result is usually stored in what's called a staging area.

Transformation is the act of cleaning up and restructuring the data to suit your needs. You might, for instance, convert your tables from third normal form to a star schema which is more suited for traditional data warehousing.

Finally, you need to load the new data into the target system in a way that keeps everything in sync. There are a lot of special cases and things to think about. To do all this well, you need to use several tools in the toolbox.

In many cases, special ETL tools are used to handle all these operations. You might have come across software from companies like Informatica, IBM, and Oracle just to mention a few. A large number of open source alternatives are also available.

As you might have gathered by now, pretty much all the stuff you can do in these graphical tools you can do from the prompt as well. Reading, manipulating, and writing data to Databricks are a breeze as we've seen in the last few chapters.

The thing is that there are a lot of things you can do using code that isn't available in the graphical tools. Let's dig even deeper into ways of cleaning up data and learn a few tricks along the way.

An overview of the Spark UI

One thing we haven't talked about before is the Spark UI. It's a number of screens that provide you with a graphical representation of what's happening in the system. Using this, you can get more information about what's happening in the system when you ask Apache Spark to run a job for you.

You can get to this screen via the cluster detail page. In reality, you'll most frequently open it from the notebooks. Every time you run a command on the cluster, you'll get a Spark Jobs line that you can expand.

If you do that once, you'll get a list of jobs and a view link. Click that, and you'll open the Spark UI on the Jobs page. This view will show you the job status and the stages. You also get a couple of visualization options.

The event timeline will show you how the work is going over time and what's actually being done. This can be useful in many cases, for instance, if you think you have skewed data or unbalanced work in any other way. You want the work to be as evenly distributed as possible. A nightmare scenario would be to have all but one executor finish in five minutes and then have one straggler keep working for a long time.

We already mentioned the Directed Acyclic Graph, or the DAG. This is where you can actually see the work that is being done in the order Apache Spark is working on it – useful for work optimization. This is where you'll find that work is not being done in the right order.

If you go back to the notebook and drill one step further down in the Spark Jobs hierarchy, you'll see the stages. To the right of them, there's a small information icon. That will take you to the stages view for the stage you pick. You'll get the same view as for the job, metrics, and a task list.

From these pages, you can look at even more details by clicking the tabs at the top. Storage will give you information about what's happening on the I/O level, including data about how well the cache is working.

The Environment tab will just show you all the current settings in the cluster. Executors will present you with data on what's happening on the system, and SQL will show you information about the query if what you're running in the cell is indeed SQL.

A good way of getting to find your way around the Spark UI is to open it in the notebook, run queries, and click the reload button in the top-right corner of the page. That'll update the information. Let's try a simple one:

```
df = spark.read.csv('/databricks-datasets/airlines/part*')
df.count()
```

If you open the first job view while it's running, you'll find there's one active stage that is listing leaf files and directories. It's looking at all the files in the folder. Look at the stage details, and you'll see there are a ton of tasks. Some are probably done by the time you look at them.

The next stage is globbing paths and checking file existence – making sure files are there, basically. Then it's time to actually do the transformation df = spark... If you look at duration, you'll notice this step takes almost no time. It's a transformation after all.

Finally, it's time to run the count, which requires all the files to be read. If you look at the event timeline for the stage, you'll see that there are a lot of mostly equally sized boxes being read across all nodes. Processor cores are staring to spin for real. Click the Executors tab, and you will see how the load is distributed among the nodes. Once they're done, the number is shown to you.

This is a simple example of how you can track what's going on and make sure the system behaves as you expect it to. Looking at these views frequently is good. Getting a gut feeling for is crucial if you want to do good work. You'll save a lot of time if you can spot that something is off early on and not after 100 hours of erroneous data crunching.

CHAPTER 8   ETL AND ADVANCED DATA WRANGLING

Cleaning and transforming data

With all the hype around data science, you might think that most work in the field is about developing cool algorithms and applying them while chanting magical spells, like a data Gandalf or Harry Potter.

Not so much, unfortunately. Most work in the data analysis field is still about understanding the data and preparing it for consumption. It doesn't matter much if it's humans or algorithms that will use it eventually, it needs to be put into neat tables.

There are tons and tons of issues that you'll have to handle along the way. Dealing with bad data, null values, and erroneous metadata are some examples. Then you usually have to change the format, structure, or content in some way.

Let's create a DataFrame with errors in it to see how we can get rid of them. To do that, we use the createDataFrame procedure. The first list contains the data, the second one defines the schema or (like in this case) just a bunch of column names. In the latter case, the actual schema will be inferred:

```
df = spark.createDataFrame(
 [
  ('Store 1',1,448),
  ('Store 1',2,None),
  ('Store 1',3,499),
  ('Store 1',44,432),
  (None,None,None),
  ('Store 2',1,355),
  ('Store 2',1,355),
  ('Store 2',None,345),
  ('Store 2',3,387),
  ('Store 2',4,312)
 ],
 ['Store','WeekInMonth','Revenue']
)
```

Note that we have a couple of errors here. We have a few None values, which is what Python uses for Null. Then there's a bad value for the first store's fourth row. There are no months with 44 weeks as far as I know. Also, there's a duplicate in there.

Finding nulls

First, let's see how we can identify the different issues. We'll start by looking for null values. There's a nice little function for that called isNull. Filter using that, and you'll get all the rows containing null values. Let's try it:

```
display(df.filter(df.Revenue.isNull()))
```

This is nice, and in this example we could do this for all three columns to look for data. If you have 100 columns, it's not as fun though. Then it might be better to start with looking at the number of nulls for each column. There's a number of ways to do this, but let's do it using SQL functions:

```
from pyspark.sql.functions import count, when, isnull
display(df.select(
  [count(when(isnull(c), c)).alias(c) for c in df.columns]
))
```

To start with, we import the functions we need. Isnull is needed to identify the none values we're looking for. When is a conditional check, and finally count helps keep the tally. We then loop through the columns and count the number of nulls in each of them. To keep the column names for the aggregates, we use the alias command:

```
from pyspark.sql.functions import col

cols = [c for c in df.columns if df.filter(col(c).isNull()).count() > 0 ]
```

This is an alternative take on the same problem. This will return all the columns where there is at least one null value. The cols variable will contain the list so that you can do logic or just run a df.select(cols).

If there are not a huge number of nulls, you'll probably want to look at the actual rows with the missing data. While you could create a query where you manually add filters for each column in the cols list, you can also just have Python do it for you, with the help of reduce:

```
from functools import import reduce

display(df.filter(reduce(lambda a1, a2: a1 | a2, (col(c).isNull() for c in cols))))
```

143

CHAPTER 8 ETL AND ADVANCED DATA WRANGLING

This time we import yet another function. Reduce executes a function for each value in a list. For instance, you can create a function that adds to numbers; use reduce on a list like 1, 2, 3, 4 and it'd return 10.

In our case, the result of the call will be ((Day IS NULL) OR (Revenue IS NULL)). This is then passed back to filter, which will execute the statement and return the rows that match our argument. You can of course run this across all columns directly:

```
display(df
        .filter(
                reduce(lambda a1, a2: a1 | a2, (col(c).isNull() for c in
                df.columns))
        )
)
```

You will also notice we're using something called lambda. This is basically a function that we haven't predefined. We'll come back to that later in the chapter, so don't worry if that part is a bit confusing. It'll clear up in a bit.

Now you know where the null values are. The question is what to do with the values once you've found them. Ideally you can go back to the source and get the data. Oftentimes the data is not there either. So you'll have to resort to do some clever engineering.

Getting rid of nulls

You have a number of programmatic options. None of them are great though. They can't be as you're either throwing away data or guessing values in some way. You basically have to pick your poison based on your data knowledge. Let's look at a few tools in your toolbox.

The easiest option is to just remove the rows with the null values. While deleting data never feels good, it can in many cases be a quick and valid solution. If you, for instance, have 100 million rows, it probably doesn't matter much if you lose out on a few hundred rows. The result will in most cases be close enough.

The command for this hard-core row removal is dropna. Let's see what happens when you just run the command straight off. This will remove any row that contains at least one field that is null:

```
df2 = df.dropna()
display(df2)
```

CHAPTER 8   ETL AND ADVANCED DATA WRANGLING

Note that this command, like many others, won't work in place. Instead it returns a DataFrame. You have to assign the result back to the same DataFrame or a new one if you want to keep both versions. If you just run dropna without assignment, nothing will happen.

As you'll notice, we've lost a few rows in df2. The command worked pretty much as expected. This is a very easy way to get rid of rows that can be troublesome. If there are just a few rows in a big dataset, this might be a valid option.

If you want to clear out only totally empty rows, you can do that using dropna as well. This sometimes happens when exporting tools add line breaks at a given number of rows. To clear them, just set the first argument to "all". Otherwise, the default is "any":

```
df2 = df.dropna('all')
display(df2)
```

If you don't want to remove rows in case there are nulls in a given column, you can limit dropna to only look at a subset. This command will remove rows where there are null values in the two columns specified, but leave nulls if they are in the Revenue column:

```
df2 = df.dropna(how='any', subset=['Store','WeekInMonth'])
display(df2)
```

There is another parameter that can sometimes be useful. If you have a lot of columns, you can specify how many valid column values there, at least, have to be to keep them. It's called thresh. If you use it, it doesn't matter what you type in – "any" or "all" – for the first argument as it'll be ignored:

```
display(df.dropna(thresh = 2))
display(df.dropna(thresh = 3))
display(df.dropna(thresh = 4))
```

With the first command, we only got rid of the row with all nulls. The reason is we have at least two fields populated in the remaining rows. When we bump up the requirement to three, all our nulls are gone. The next one is interesting. It'll actually remove all rows as there are only three columns in our table. Beware of this little feature.

Filling nulls with values

Next up, let's look at what you can do if you don't want to remove any valid data. The fully empty row we don't need, but the rest can be somewhat interesting. In many cases, you'll want to build your own logic. You don't have to though. There are a few standard options. One simple option is to use a fixed value using fillna:

```
display(df.fillna(0))
display(df.fillna(0, ['Revenue']))
display(df.fillna({'WeekInMonth' : 2, 'Revenue' : 3}))
```

The first example will just replace the nulls across all numeric columns with zeros. Next up, we only replace nulls in the Revenue column, leaving the rest of the columns as is. Finally, we specify the changes on a per-column basis.

There is one important thing to note about fillna – it'll only do the exchange operation for matching column types. So if you use a numeric value for a string column or the other way around, it won't work. Let's try it using the fillna command with a string argument instead of a number:

```
display(df.fillna('X'))
```

You'll notice that nothing happened with some of the null values. The reason is that they are stored in numeric columns. Beware of this as you won't get any kind of result back that indicates the result. It also means you'll have to run the command once per column type.

One common thing you might want to do is use the median or mean for a given group. This makes sense in many cases where you have a rather small spread. Sales for a large store probably won't swing back and forth a lot from one Monday to the next:

```
display(df.groupBy('Store').avg('Revenue'))
```

This will give you the values you need for your Revenue nulls. While you can do this using the commands we've already looked at, there's another alternative. That's something called Imputer. It's pretty much built for this exact thing – completing missing values. Let's look at how that works:

```
from pyspark.ml.feature import Imputer

df2 = df.withColumn('RevenueD', df.Revenue.cast('double'))
```

```
imputer = Imputer(
    inputCols=['RevenueD'],
    outputCols=['RevenueD'],
    strategy='median'
)
display(imputer
        .fit(df2.filter(df2.Store == 'Store 1'))
        .transform(df2))
```

We import the feature we need from the machine learning module. As the Imputer requires a double or float to work, we then create a new DataFrame with a new column, RevenueD. As you can see, we cast it into double.

Next, we set the parameters we need. The inputCols defines what it should look at, and outputCols says where the result goes. Note that they don't have to be the same. You can type in something like RevenueDI, which will give you a new column.

Finally, we apply our rules to the df2 and transform the same DataFrame as well. The fitting is done with a filter. The reason is that we don't want the median to be calculated over the whole dataset.

Notice also that we're using median as the strategy here. The default is mean, and you can type it in instead if you want to. What method you choose depends on the actual use case and data.

One thing worth mentioning is that you can look at the result of the fitting. You can try this for the previous code snippet, for instance. It'll show you what values it calculated based on the input. The surrogateDF is a helpful DataFrame offered by the Imputer class:

```
i = imputer.fit(df2.filter(df2.Store == 'Store 1'))
display(i.surrogateDF)
```

As you probably realize, it's the transformation part that actually applies the result to your target DataFrame. You can use the preceding fitting variable to send the result to a DataFrame, of course:

```
df2 = i.transform(df2)
```

That's enough of nulls. It was a lot, but it's a common issue so you might as well get used to working with them. As you've seen, there are quite a few ways to deal with them, and there are even more if you need to dig really deep.

Removing duplicates

The next problem was around the duplicated row. This is not a topic to dwell on. There's a very simple command that takes care of it for you. Just like dropna and fillna, it's part of the DataFrame class. So let's try it out:

```
display(df.dropDuplicates())
```

This will look at all the columns in the table and remove rows that are identical across all columns. If you want to compare on a subset of the columns, like if you have a primary or unique key in the source, you just need to add a parameter:

```
display(df.dropDuplicates(['Store','WeekInMonth']))
```

This is an easy function to use, but it doesn't really give you any way to decide which row to keep. This is a rather common issue that you will run into sooner or later. For instance, imagine a replacement for a row comes in before you processed the original one. If you only need to handle the newest, you want to get rid of the old one. Normal duplication handling won't solve this consistently. Let's look at an example:

```
display(df.dropDuplicates(['Store']))
```

This should give you the first row for each store. While you can coalesce and try to flip data around to get the result you want, it's not consistent. Also, collecting everything into one partition is terribly inefficient. Instead, you need to use another technique.

You might remember that we've talked about analytic functions and the power you get by using them. While we will talk more about them in a later chapter, this is a good use case for them. Let's see how they can help:

```
from pyspark.sql.window import Window
from pyspark.sql.functions import row_number, desc

w = Window
    .partitionBy(df['Store']).orderBy(df['WeekInMonth'].desc())
display(df
    .select(df['Store'], df['WeekInMonth'],
        row_number().over(w).alias('Temp'))
    .filter(col('Temp') == 1))
```

There's a lot happening here, so let's go through it. I guess you're getting familiar with the import parts. We're just getting more functions we need from pyspark.sql. Window will enable us to use the windowing functions in Python, and row_number will help us order the rows. Desc, short for descending, will flip the sorting order.

Then we create a window by partitioning the data by the Store column. We then order each partition by the WeekInMonth column. Desc gives us the highest number at the top. The window is stored into a variable.

Next up, we do a normal select on our DataFrame with Store and WeekInMonth to start with. It's the next part that does the magic. The function row_number orders the numbers in our window from 1 to n. It will do it per partition. The resulting column is named Temp by the alias command.

We then filter this result and only pick up the top values according to our sorting order. As the first value is always 1, we can just look for that number. Using this method, you can choose what rows to keep in a duplicate situation.

If the windowing seems confusing, try running the command without the filter. That way you'll see what the Temp column looks like. You might also want to revisit Chapter 6 for more information.

There is one thing that is rather sad with this example though. We end up with an unnecessary column that we don't want. Let's see if there is anything we can use to get rid of it, like a drop command, maybe:

```
display(df.select(df.Store, df.Day, row_number()
 .over(w).alias('Temp'))
 .filter(col('Temp') == 1)
 .drop('Temp'))
```

There we go. A simple drop command removes the column. As we put it last in the chain, it will be removed after it served its purpose. You can of course use this for normal columns as well, as we've mentioned in an earlier chapter, like this:

```
display(df.drop('Revenue'))
display(df.drop('Revenue','Store'))
```

CHAPTER 8   ETL AND ADVANCED DATA WRANGLING

Identifying and clearing out extreme values

The last issue in our dataset is the extreme value. Like with anything we've talked about here, there are a lot of ways to do this. Some are simple, some are more complex. The best way is to look for anomalies.

One common solution is to look at means, standard deviations, and extreme values. If you have a lot of historic data, you can look at the patterns that exist in the dataset and check if the new information looks similar.

To do this, you need to get data about your data. There are many ways to do that. If you run stuff manually, there's a little trick you can use: graphs. The built-in plot engine is a pretty nice tool for finding basic issues like the one we have here. Run this basic query to start with and you'll see:

```
display(df)
```

So this just gives us the data. Now click the plot button, and you'll get a weird bar chart. Click the Plot Options button. In the new view, add Store to Keys and Day to Values. Set aggregation to count and finally choose Histogram plot as the display type. Apply and you'll see two charts with dots representing the days.

For the first store, you can see that the data is clustered on the left-hand side with an outlier on the far end to the right. As we don't have much data, it almost looks as if it's just a strange distribution, but add another few correct rows and this will shine even brighter.

If you move your cursor over the bar, you'll see the value. This helps you locate the outlier, making it easier to fix the issue. Other charts, like boxplots and q-q plots, can be helpful as well in the same way.

This method isn't very useful if you want to test things programmatically though. For that you need a different way. A simple one is to use the describe function. Let's look at what information you get back if you use that one on our dataset:

```
display(df.describe())
```

You get a lot of goodies out of this simple command: count, mean, standard deviation, min value, and max value for all the columns for which the values make sense. You can of course filter DataFrame to get the data for a subset:

```
display(df.filter(df.Store == 'Store 1').describe())
```

This is the easiest by far and will get you a bunch of information. It is, however, a bit unnecessary if you just want one value. There's no need to calculate everything if that happens to be the case.

If you want to do it more directly, there are functions you can just call to get the same results as in the preceding code. Let's look at an example for pulling out the mean and standard deviation for just Revenue. Also, let's group it by Store:

```
from pyspark.sql.functions import mean, stddev

display(df
       .groupBy('Store')
       .agg(mean('Revenue'), stddev('Revenue')))
```

That works great. You can now do a t-test if you think that makes sense. While there are a lot of other mathematical functions in the SQL library, you won't find median. For that you'll have to do some math on your own. I personally think the easiest way is to use a function called approxQuantile:

```
df.approxQuantile('Revenue', [0.5], 0)
```

This will give us the value at a given quantile, in the span from 0 to 1. So if you set the second argument to 0.0, you'll get the lowest value for the column. With 1.0 you get the highest value. In the middle you get the median, which is what we are looking for.

The last argument is interesting. It lets you define the relative error, from 0 to 1 with 0 being no error. In this case we set it to 0 to get the exact value. If you have a lot of data though, you might be better off doing a rough estimate. That will make approxQuantile sample a subset instead of running a calculation on the whole column.

If you want to flag odd values, you can create a new column and populate it based on one or more filters. There are other ways though. Let's have a little fun and at the same time introduce yet another tool to your expanding toolbox:

```
from pyspark.ml.feature import Bucketizer

mean_stddev = df
  .filter(df.Store == 'Store 1')
  .groupBy('Store')
  .agg(mean('WeekInMonth').alias('M'), stddev('WeekInMonth').alias('SD'))
  .select('M','SD')
  .collect()[0]
```

```
mean = mean_stddev['M']
stdev = mean_stddev['SD']

mini = max((mean - stdev),0)
maxi = mean + stdev

b = Bucketizer(splits = [ mini, mean, maxi, float('inf') ]
            ,inputCol = 'WeekInMonth'
            ,outputCol = "Bin")

dfb = b.transform(df
                    .select('WeekInMonth')
                    .filter(df['Store'] == 'Store 1'))
display(dfb)
```

So let's see. At the top we start by getting the mean and standard deviation for the Day column for Store 1. We return the data to a list that we then use to create min and max values for one standard deviation. These values are then used to populate the Bucketizer class.

You might be familiar with the concept of buckets or histograms. We basically split our data into a given number of slices, or bins. We'll get back to this idea when we talk about partitions, but in this case we use it to identify outliers.

The splits argument is defining the buckets. We have four here, but you can have as many as you want: the three we just built and an infinite one to capture extreme values. Next, we define the input and output columns. The output one is a new one in this case. It'll be created.

Finally, we do the actual transformation. The result is the data we're used to, but with a new column at the end. In that one you'll see which bucket the row belongs to according to the rules we defined.

As you'll see the three first values end up in the first bucket. It's a zero as that's what the index starts at. These three values are higher than our defined min value, but lower than the value for the first bucket. Our outlier ends up in the third bucket:

```
display(dfb.groupBy('Bin').count())
```

This will give you a quick view of the distribution. While distribution is no silver bullet by any stretch of the imagination, it can sometimes be helpful. It's especially good at capturing extreme values. Mistakes like someone forgetting to use the decimal point on a cash register can sometimes be captured this way.

Taking care of columns

With our data issues out of the way, it's time to look at other things you might bump into - things like changing names of columns. If you read data from a CSV file, for instance, you might end up with strange headers. In many cases they can hinder you from creating a table in Databricks:

```
df = spark.createDataFrame(
[
  ('Robert',99)
],
['Me, Myself & I','Problem %']
)

df.write.saveAsTable('Mytab')
```

If you run this code, you'll get an error telling you that the columns contain invalid characters and that you should rename them. This is of course possible to do manually one by one, but not very efficient if you have a lot of strange columns. Let's investigate an alternative route that might be faster and easier:

```
newnames = []
for c in df.columns:
  tmp = c
      .replace(',','-')
      .replace('%','Pct')
      .replace('&','And')
      .replace(' ','')
  newnames.append(tmp)

display(df.toDF(*newnames))
```

We loop through the columns and just replace the list of characters we don't want with decent alternatives. Notice the last row. To set the new names, we use the toDF with a prepared list. The asterisk unpacks the list so it can be passed to the procedure. There are many other ways to set new names, and we've seen a few, but I like this one.

An even faster alternative, if you don't want to control the name change in detail, is to use regular expressions. While it's much less work, remember that this can actually fail if you have two columns with the same name plus a special character. For instance, "Revenue" and "Revenue %" would both end up like "Revenue":

```
import re
newnames = []
for c in df.columns:
  tmp = re.sub('[^A-Za-z0-9]+', '', c)
  newnames.append(tmp)

display(df.toDF(*newnames))
```

If you're not familiar with regular expressions, I recommend reading about it, for instance, in *Regex Quick Syntax Reference* (www.apress.com/gp/book/9781484238752). Nothing will make you feel like a data ninja like knowing your way around it. Anyways, in this case we look for everything that's not letters and numbers. It's the caret sign that negates the match. We replace all non-wanted occurrences with nothing.

Pivoting

Sometimes you want to flip your data from rows to columns. You might do this all the time in Excel. The feature there is called pivoting. The same thing is available in Pyspark. It's even called the same thing, which is helpful.

You're basically rotating the data around a given axel, hence the name. In this case that axel is the data in one of your columns. Let's look at an example that might make this easier to understand, if you're new to the concept:

```
df_pivoted = df
        .groupBy('WeekInMonth')
        .pivot('Store')
        .sum('Revenue')
        .orderBy('WeekInMonth')

        display(df_pivoted)
```

We group by the WeekInMonth column and sum the revenue. So far you recognize what we've already done. What are added are the pivot functions. They'll pull out the values in the Store column and use them as column headers.

This will give you a new DataFrame with the revenue shown in the intersection of each store and week. While you can show the same data row-wise, this often creates a clearer view. Feel free to compare yourself with this snippet – same data, but much harder to read:

```
display(df
        .groupBy('Store','WeekInMonth')
        .sum('Revenue')
        .orderBy('WeekInMonth'))
```

While pivoting is useful for visual clarity, it has many other uses. For instance, it's something you'll frequently use when preparing data for machine learning use cases – yet another tool in the toolbox.

If you want to go the other direction, it's possible, but not as easy. There is no built-in solution for this. We need to use a bit of trickery. Let's look at what that can look like for our preceding example:

```
display(df_pivoted
        .withColumnRenamed('Store 1','Store1')
        .withColumnRenamed('Store 2','Store2')
        .selectExpr('WeekInMonth',
            "stack(2, 'Store 1', Store1, 'Store 2', Store2) as (Store, Revenue)"))
```

In this example, we start by renaming the column names. The reason is we need to reference them later in a context where spaces are not allowed. We then run selectExpr, which accepts logic within the select statement.

The trick here is the stack command, which is a table-generating function that returns the data in the order we specify it. The first value specifies how many columns we want. Then we set the row name, the row value, the next name, and so on. Finally, we set the new column names.

While the stack command is useful for un-pivoting data, you can use it on its own as well for creating tables. So if you quickly want to whip up some data, you don't have to use the createDataframe command:

CHAPTER 8   ETL AND ADVANCED DATA WRANGLING

```
df = spark.sql("select stack(3,'Store 1',1, 448,'Store 1',2, 449,'Store
1',3, 450 ) as (Store,WeekInMonth,Revenue)")
```

display(df)

Explode

There's another situation that you'll come across from time to time. Multiple data points sometimes arrive in one column. This is especially true when JSON is the source format and the information was just pushed into a relational database.

You can solve this issue by using the explode command. It'll take the string with multiple values and place them on a row each. A simple example will show you what that looks like:

```
from pyspark.sql.functions import explode

df = spark.createDataFrame([
        (1, ['Rolex','Patek','Jaeger']),
        (2, ['Omega','Heuer']),
        (3, ['Swatch','Rolex'])],
        ('id','watches'))
```

display(df.withColumn('watches',explode(df.watches)))

We start by creating a new DataFrame. As you can see, I add multiple values to the same column. While this is convenient when adding data, it's a bit messy when reading it. That's why we use the explode command to create a more structured table.

What's actually happening is that Apache Spark creates a new row for every string in the specified column. All data from other columns is copied over to the new rows. Now you can run aggregations like you're used to.

Beware that this denormalizes your dataset. As mentioned earlier, that is not a bad thing, but you need to be aware of it. Otherwise, you might do mistakes when you run aggregations on the information.

When being lazy is good

There is one more thing worth mentioning here. We've talked multiple times already about how lazy Pyspark is. This can be used both to your advantage and to your detriment. Stacking logic is good. This makes it possible for the optimizer to collect everything at once and create one, overarching plan on how to get data, for instance:

```
df = spark.sql('select * from Mytab')
df.filter(df['mycol'] == 'X')
df.count()
```

The first two commands aren't actions. This means they won't do the heavy load of actually picking up the data. The last one will, however. The great thing here is that you'll only read the filtered data, even though you asked for all of it initially. So being lazy helps in this case.

There are cases where being lazy is not great though. One example is when you need to use a DataFrame several times and there is a bit of processing each time. This can be the case when you do joins, for instance:

```
df_huge = spark.sql('select * from hugetable')
df_bigdim = spark.sql('select * from bigtable')
df_combined = df_huge.join(df_bigdim,'commonkey')
```

Now, if you plan on doing some kind of logic that requires iteration over the dimension table, being lazy is hurtful. The reason is you'll redo the join every time that you call the df_combined DataFrame. You can of course add filters, but it's frequently better to materialize the query by creating a temporary table. Note that a view will not help you here. Data needs to be stored.

This is part of creating flat tables to begin with. It is called denormalization and undoes the normalization you normally use in traditional databases. The idea is to accept redundant data, increased storage use, and a bit of up-front processing as to speed up read access.

If you're storing data in the Parquet format, this technique will work well as you'll only read the columns you need. You can add redundant attributes at the end of your table without taking a performance hit.

It will consume more storage. As disk isn't very expensive anymore, this isn't something you should be afraid of. If you can avoid frequent joins, you'll save a lot of time and resources in the future.

If you have to do joins, there are a few things to think about. The way Apache Spark works, it really doesn't like huge joins. With data being spread across many nodes, it requires a bit of shuffling that is a heavy burden.

Ideally one side of the join is a small table, like a dimension table. If so, Apache Spark will distribute a copy to every node. This is called a broadcast join. Normally this will happen under the surface, but if you want to make sure this is happening, you can:

```
from pyspark.sql.functions import broadcast
df_result = df_huge.join(broadcast(df_bigdim), 'commonkey').explain()
```

This will force the distribution of the dimension table. The explain command returns the optimizer reasoning. You'll find that the data from bigdim is sending data (BroadcastExchange) and that a BroadcastHashJoin is done. We'll go through how to read these plans in the advanced section.

If you remove the broadcast command and run explain, you'll most likely find that Apache Spark does this anyway. At least if the dimension table really is small. Trying to broadcast a large table is not a great idea. Try to flip the preceding command, and it'll likely fail due to size.

A few other tricks are also available. Like with the preceding broadcast join, this is not something you should have to deal with as most of this is handled under the surface. There are situations that might make them relevant though. So let's have a brief discussion about them.

Caching data

One thing you'll probably come in contact with pretty quickly in Databricks is caching. Unless you turn suggestions off, you'll see this one pop up when you run queries on the same table. The idea is that you should give the nodes faster access to data by storing it locally in memory. By doing this future queries on the same table should go faster. This is how you force a table to the cache:

```
CACHE TABLE mytab;
```

If you don't want to read the data immediately, you can add the keyword lazy. This will tell Databricks you want this table cached, but not necessarily right now. Instead it'll happen as you select data in other queries:

```
CACHE LAZY TABLE mytab;
```

When you don't need the table anymore, it's usually a good idea to get rid of it from the cache. Otherwise, it'll just consume resources unnecessarily. That said, Apache

Spark will eventually clear it out using a least recently used, LRU, algorithm. This is important to remember if you don't feel that the cache is working. It might very well be you overused the resources and the table was pushed out. Anyway, this is how you remove a table from the cache:

```
UNCACHE TABLE mytab;
```

Of course you can do this in Python as well. The commands are cache and unpersist. The result is the same as if you do it in SQL. These examples assume you have the DataFrame loaded, of course:

```
df.cache()
df.unpersist()
```

The preceding examples are using Apache Spark caching, and it's always available. On top of that, you can activate and use the Delta cache. This is a Databricks add-on that gives you access to a new type of caching. The core idea is the same as in general Apache Spark caching, but they do it differently.

One important example is it's disk based instead of memory based. While this might sound like a worse solution, the truth is that modern storage systems using solid-state drivers are really quick. Memory is of course faster, but it's also scarcer.

Another big difference is that caching happens automatically. You don't need to specify what data needs to be cached. It just happens, if the feature is enabled. If you want, you can still add objects explicitly. Plus you can cache parts of tables, as we'll see.

Note that even though the name implies that you must use Delta Lake, it actually works on any Parquet file. It does not work with any other file type though. It doesn't have to be stored on the Databricks File System, but only Parquet is supported.

To use Delta cache, you can pick one of the preconfigured worker node types. Look under the storage optimized selection, and you'll find options with Data Cache Accelerated within parentheses. Using these you'll get the configuration settings you want for this feature.

If you want to set it up manually, you need to set a few flags. You can do it in the cluster configuration or using commands. To see your current settings, you can run the following commands. They will, in turn, tell you if the feature is enabled, if data in the cache should be compressed, how much disk is allowed to be used for caching on each worker node, and finally how much space can be used for metadata:

```
print(spark.conf.get("spark.databricks.io.cache.enabled"))
print(spark.conf.get("spark.databricks.io.cache.compression.enabled"))
print(spark.conf.get("spark.databricks.io.cache.maxDiskUsage"))
print(spark.conf.get("spark.databricks.io.cache.maxMetaDataCache"))
```

None of the last three parameters matter if you don't enable the feature, of course. You can enable the Delta cache by setting the spark.databricks.io.cache.enabled flag. You do it like this, in a Python cell:

```
spark.conf.set("spark.databricks.io.cache.enabled", "True")
```

For the other parameters, it makes sense to set it up so that you don't use up more than half of the available disk space on the worker node. That is also what you'll get if you use one of the preconfigured node types.

As mentioned you don't need to explicitly cache data. If you however want to cache part of a table, you can do that explicitly. This might make sense if you have a lot of history data in your table and just want to keep the last month at hand:

```
CACHE SELECT * FROM mytab WHERE year = 2020;
```

That's pretty intuitive. You just write your query and add the cache command in front. As mentioned you shouldn't have to do this explicitly. If you ask for that data enough, Databricks will figure it out for you.

You might think caching sounds great and that everything should be cached. Well, that's not a great idea. The thing is that Databricks is pretty good at shuffling data anyways. Plus you hopefully aren't revisiting the same data too frequently. Using up local memory and storage might be a waste of time and resources.

On the other hand, if you know you'll be using the same dataset over and over, you might find that caching helps. Like with any tool, you need to look at your particular use case and see if it makes sense.

Data compression

You're probably familiar with compression generally. It's a way to shrink data. This saves space both on disk and in memory, so it's usually a good thing to do. There is, however, a drawback; and that's the decompression part. It takes a bit of processing power to take the data back to the state you need it in for analysis work. While data analytics is typically not CPU bound, it's good to keep this in the back of your mind.

In some cases, like music and video, you can accept some loss information. This is not the case for analytics, unless you work with images and sound. For traditional data you need lossless compression.

There are a lot of different algorithms that try to do this in the best way possible. As none of them can do too crazy things, they usually balance speed and compression rate. Faster usually means you gain less compression.

When you use the Parquet file format, you'll by default use the snappy algorithm. That provides you with a good balance of compression ratio and speed. There are other options though. Depending on your data and sensitivity to speed, you might find that other alternatives are better for you.

The options you get out of the box are no compression, gzip, lz4, and the aforementioned snappy. Let's loop through the different options and see the differences in a smallish dataset:

```python
import time

targetDir = '/filetemp'
dbutils.fs.mkdirs(targetDir)

sourceData = '/databricks-datasets/power-plant/data'
files = dbutils.fs.ls(sourceData)
originalSize = sum([f.size for f in files])

df = spark
  .read
  .options(header='True'
          ,inferSchema='True'
          ,delimiter='\t')
  .csv('{}/*'.format(sourceData))
algos = ['uncompressed', 'lz4', 'gzip', 'snappy']

rounds = 3
for a in algos:
  print(a)
  runTime = 0
  for i in range(rounds):
    startTime = time.time()
```

```
    df.write
      .mode('overwrite')
      .parquet('{}/{}.parquet'.format(targetDir,a), compression = a)
    runTime = runTime + (time.time() - startTime)
    print('Run {} {}'.format(i, time.time() - startTime))
  print('Avg {}'.format(runTime/rounds))
  files = dbutils.fs.ls('{}/{}.parquet'.format(targetDir,a))
  compressedSize = sum([f.size for f in files])
  print('Size {}'.format(compressedSize))
  print('Gain {}'.format(round(1-(compressedSize/originalSize), 3)))

dbutils.fs.rm(targetDir, recurse=True)
```

We start by creating a temporary folder on the Databricks File System. Then we read some of the standard data you have available. We measure the size of the files in there to get a baseline for our compression efficiency check.

The algos variable lists the four options we'll try here. With rounds we define how many tests we do for each compression algorithm. While I typed in three here, feel free to add a few rounds.

In the actual loop, we write to a file, named after the algorithm we're using. We show the runtime for each loop and also the average. Next up, we loop through the files in the folder and sum them up to see how well they perform compared to each other. With these numbers we calculate how much space we saved.

In my case, gzip gave me the best compression, while snappy was fastest. The exact number for execution time will vary, of course. The limited size will not give a true value for speed, but the ratio will be the same every time. Note that you get quite a deal of compression even without using any algorithm. Text files simply aren't very efficient for data storage.

If you want to test the same thing using text files, you can do so. Just change the Parquet parts to CSV. In this case you'll notice that uncompressed really means what it says. It'll just rewrite the file:

```
.csv('{}/{}.csv'.format(targetDir,a), compression = a)
files = dbutils.fs.ls('{}/{}.csv.format(targetDir,a))
```

The last thing we do, by the way, is to clean up after ourselves. If you want to keep the data and look at the results more closely, just comment that row. Just remember to clean it up later in that case.

If you want to change the compression code you're using on a cluster-wide level, you can set a flag to achieve just that. This will change the format of all the Parquet files you store without the compression flag. The following second command will show you the current default:

```
sqlContext.setConf('spark.sql.parquet.compression.codec', 'gzip')
sqlContext.getConf('spark.sql.parquet.compression.codec')
```

Note by the way that measuring clock time is not a great way to evaluate performance as there are many external things that can affect the performance. That's especially true if the things you measure take a fraction of a second to run. In those cases it's better to use something like timeit:

```
%timeit sum([i for i in range(10000)])
```

If you really want to go the extra mile, you could look at another algorithm. Lempel-Ziv-Oberhumer, or LZO, can sometimes generate better results than snappy. It is, however, not preinstalled on Databricks. This apparently has to do with licensing. You can install it yourself if you want. Look in the documentation if you want a guide.

A short note about functions

Throughout this book and chapter, we've been using a lot of different modules that offer a lot of extended functionality. Sometimes you want to do the same thing in your own code. Instead of rewriting the same lines over and over, you can create your own functions.

We've covered this a little bit earlier, but now it's time to give them another visit. We need to talk a bit about vectorized UDFs and lambda functions. This time we'll just look at Python so if you're interested in the SQL ones, revisit Chapter 6 for an example of the syntax.

It should be noted that while user-defined functions, or UDFs, have their place, you shouldn't be using them unless you really have a use for them. It's easy to kill performance by creating a bad one in Python.

You should also not try to recreate something that's readily available. While it's easy to write a working linear regression from scratch, you probably won't create one as good as the one offered in the MLlib package. So use that one instead. Most general things are already created.

Lambda functions

While you usually create functions so that you can reference the same code over and over, there are situations where you just want a one-off quick-and-dirty solution to an issue. This is where lambda, sometimes referred to as anonymous, functions come in. They work like any function, you just don't need to declare them. You might recall that we already used them in this chapter:

```
display(df.filter(reduce(lambda a1, a2: a1 | a2, (col(c).isNull() for c in cols))))
```

Let's look at an example of a normal function and an anonymous one. Both do the exact same thing. The difference is that the first one is predefined, while the other one can be created in-line:

```
def plusOne(i):
        return i + 1

lambda i: i + 1
```

These examples just define what the function does. It doesn't actually run anything. If you want to execute the code, you have to do a little bit more. The easiest way is to just wrap the lambda function into parentheses and send an argument:

```
(lambda i: i + 1)(1)
(lambda i, j: i + j)(1,2)
```

This isn't super useful though. Where lambda functions really add something is like in the example early in this chapter. They also shine together with another neat feature: map functions. Map functions let you iterate over dictionaries, lists, and so on and call a function for each item. This is convenient if you want to keep your code concise:

```
countries = ['Sweden','Finland','Norway','Denmark']
mapResult = map(lambda c: c[::-1], countries)
print(list(mapResult))
```

This code will take the countries in the list, flip them over, and return the results. The lambda function is the one doing the reversal using string slicing. The -1 in there tells Python to walk the string backward.

The countries are passed on as parameters and sent into the lambda function. Everything is sent back in the form of a map object. We convert that into a list and print it out so we see the results.

A detail is that you can actually name lambda functions. That makes them pretty similar to normal functions. Here's a simple example that shows you how to do it. I personally don't use this much, but you should know it's there:

```
plusOne = lambda i: i + 1
plusOne(4)
```

If you use this a lot, consider using a normal function. While it's a little bit more cumbersome, readers of the code will have an easier time finding it. As mentioned many times before, readability is very useful.

Storing and shuffling data

We've already talked quite a bit about storing data, but why not round off this chapter by revisiting yet another topic? More specifically, let's look at partitioning, save modes, and table management.

Save modes

One thing we haven't mentioned when talking about saving data is the different save modes. You can tell Apache Spark what to do if there's already data in the place you want to use as the destination.

There are four different modes you can choose from. The first one is error. It'll throw an error if the file or folder you're writing already exists. This is the default option, as you might imagine:

```
df = spark
    .read
    .option('delimiter','\t')
```

```
        .option('header','False')
        .csv('/databricks-datasets/songs/data-001')
df.write
        .csv('/temp/data.csv')
```

We read some data with the first command and then rewrite it to a new place with the second one. This will just work as the file doesn't exist to begin with. Now try to rerun the second command:

```
df.write
        .csv('/temp/data.csv')
```

This will give you an error message telling you that the path already exists. If you want to replace the existing file, you can tell Apache Spark to overwrite the file. Let's try that using the same input data:

```
df.write
        .mode('overwrite')
        .csv('/temp/data.csv')
```

Even though the file is there, this will finish ok, replacing the old files with the new ones. Now assume that you don't want to replace the information, but rather just add a new one. There's a mode for that as well:

```
df = spark
        .read
        .option('delimiter','\t')
        .option('header','False')
        .csv('/databricks-datasets/songs/data-002')
df.write
        .mode('append')
        .csv('/temp/data.csv')
```

This will add the new data in new files. If you look at the folder, you'll see that there's data from both writes, using different IDs. Read it back in, and you'll see you get data from both datasets in one go:

```
df = spark
    .read
    .option('delimiter','\t')
    .option('header','False')
    .csv('/temp/data.csv')
```

Finally you have the ignore option. This works like the default error mode, but ignore won't throw an error. This is useful if you don't want your job flows to fail out when this error occurs:

```
df.write
    .mode('ignore')
    .csv('/temp/data.csv')
```

This will work just fine, but nothing will actually happen. If you look in the folder before and after the command, you'll notice there are no differences in the data. If you use this, make sure you remember how it works so you don't read old data by accident.

Managed vs. unmanaged tables

By default the tables you create are managed. That means that Databricks cares for both the data and the metadata. This is not necessary. If you want to, you can handle the data and just have Databricks keep track of the metadata. This is especially relevant if you want to store information on a data lake.

The main difference between these two options is what happens when you ask Databricks to remove the table. If you have a managed table, the underlying files will disappear as well. With an unmanaged table, this is not true. Let's do a small test to validate this claim:

```
df = spark
    .read
    .option('delimiter','\t')
    .option('header','False')
    .csv('/databricks-datasets/songs/data-001')
```

```
df.write.saveAsTable('songs')
dbutils.fs.ls('/user/hive/warehouse/songs')
spark.sql('drop table songs')
dbutils.fs.ls('/user/hive/warehouse/songs')
```

In this example, we read some data, write it to a table, and look at the files created. Then we drop the table and try to look at the files again. You'll notice they aren't there anymore. When you dropped the table, Databricks cleared out the files for you as well, saving you the trouble. Now let's try to do this with an unmanaged table:

```
df.write.csv('/temp/songs')
dbutils.fs.ls('/temp/songs')
df.write.option('path', '/temp/songs').saveAsTable("songs")
spark.sql('drop table songs')
dbutils.fs.ls('/temp/songs')
```

In this example we copy the data to a folder and then create an unmanaged table using the path option. Then we drop it from Hive. As you'll see the files aren't touched. The only thing you removed was the metadata reference.

There is one situation that you should be aware of. If a managed table creation is interrupted, you might end up with a messy situation on your hand. Run the following command, making sure you cancel it after it starts but before it finishes:

```
df = spark
      .read
      .option('delimiter','\t')
      .option('header','False')
      .csv('/databricks-datasets/songs/data-002')

df.write.saveAsTable('songs2')
```

If you managed to break the job in time, you'll now be in an unstable state. Try to run the second command again, and it'll fail with an error telling you that the table exists. If you however try to read the table, you get another error saying the table doesn't exist.

The reason is that the folder was created and the metadata stored before the data was completely inserted. If you end up in this situation, you'll have to resort to an unsettling option – deleting the underlying files:

```
%sh rm -r /dbfs/user/hive/warehouse/songs2
```

Notice I didn't use the %fs command. Databricks won't let you use that command to mess with the Hive metadata, as it shouldn't. This is really a last resort option that you shouldn't use unless absolutely necessary and all other options are gone.

Handling partitions

Partitions are pieces of a whole that Apache Spark can distribute between the nodes. If you want to process 10 million rows on ten nodes, you ideally would want to have 1 piece on each node. Those pieces are called partitions, and that's the level that tasks are broken down to. When you create or read data, it'll automatically be stored into partitions.

When you work with Databricks, you in most cases don't have to care about this as it's being taken care of for you under the hood. There are a number of defaults, like defaultParallelism and maxPartitionBytes, that will make sure that you get reasonable performance.

Let's take a look at what a partition really is. The easiest way is to actually look at the content. Let's read a few tunes of the songs dataset and split them into four partitions that we can then look at:

```
df = spark
      .read
      .option('delimiter','\t')
      .option('header','False')
      .csv('/databricks-datasets/songs/data-001')
dftop25 = df.select('_c4').limit(25)
dftop25.repartition(4).rdd.glom().collect()
dftop25.count()
```

The third command will return a list of rows. You'll notice that there are four lists. This is because we explicitly asked for that number in our code using the repartition command. Each list, or partition, will contain the data shown.

In the last command, we run a count. This will create four different tasks, one per partition, for processing on the executors on the nodes. One executor will count the first list, the second will take the next, and so on – in parallel, of course.

CHAPTER 8   ETL AND ADVANCED DATA WRANGLING

Even though this is being handled for you, you can end up in a bad place if Apache Spark gets it wrong. In those cases you need to repartition data to get it broken up into more manageable pieces.

There are two ways of repartitioning data. You can repartition it, which will reshuffle and redistribute the data. This is a costly operation. Alternatively you can coalesce the data, if you want to go down in number of partitions. This will just move data from the partitions you don't want to the ones you keep. Let's take a look at this:

```
df = spark
      .read
      .option('delimiter',',')
      .option('header','False')
      .csv('/databricks-datasets/airlines/part-0001*')

      df.rdd.getNumPartitions()
      df.count()
```

Like we've done so many other times, we start by pulling some data. In this case we get ten files from the airlines dataset. With the getNumPartitions command, which we can only run on the RDD, we can see that we got ten partitions.

When you run the count, expand the job and look at the first stage. You'll see that there are ten tasks, one per partition. This means that the job is spread out nicely across executors on your nodes. Let's see what happens when we change things:

```
df.repartition(5)
      .write
      .mode('overwrite')
      .saveAsTable('fivepart')

df.rdd.getNumPartitions()
```

The repartitioning command will read the data, collect into five partitions, and save the table with that setting in mind. Interestingly enough the actual DataFrame will not change as it's still pointing to the underlying files.

Notice that we go down from ten partitions to five. This is a place where we could decide to use coalesce instead of repartition. The command would look and work the same; otherwise, you'd just have to change repartition:

```
df.repartition(5)
      .write
      .mode('overwrite')
      .saveAsTable('fivepart')
```

Either way you'll end up with five partitions in the resulting table. In practice this means Databricks will create five files in the folder. You can verify this by looking at the Databricks File System using the fs command:

```
%fs ls /user/hive/warehouse/fivepart
```

This will show you the actual Parquet files that the table is made up of. It'll match the partitions we asked for. You can also look at the sizes here which might be a good thing if you don't get the expected number of partitions at read:

```
df5p = spark.sql('select * from fivepart')
df5p.rdd.getNumPartitions()
df5p.count()
```

When we re-read it to a new DataFrame, we see that the partitions are what we expect them to be. Expand the job view for the count, and you can verify that there actually are only five tasks now. Let's take this to the next step. Note though that there won't always be a one-to-one match on partitions and files. File partitioning does not dictate how the RDDs are split. Apache Spark can and will make decisions on read that are separated from the underlying file structure:

```
df5p.repartition(1)
      .write
      .mode('overwrite')
      .saveAsTable('onepart')

df1p = spark.sql('select * from onepart')
df1p.rdd.getNumPartitions()
df1p.count()
```

Same thing but with just one partition. Suddenly there's only one task to do, which means only one executor can do any work. In this case it won't matter much as the data amounts are small. With more data, there'll be an issue. If you have the capacity, feel free to test this one:

CHAPTER 8   ETL AND ADVANCED DATA WRANGLING

```
df = spark
    .read
    .option('delimiter',',')
    .option('header','False')
    .csv('/databricks-datasets/airlines/part-001*')

df.repartition(1)
        .write
        .mode('overwrite')
        .saveAsTable('onepart')

df1p = spark.sql('select * from onepart')
df1p.rdd.getNumPartitions()
```

The difference here is that it's no longer just ten files, but 100, as we've moved the asterisk one step to the left. Suddenly things take a long time to complete. As mentioned the reason is that there's now one core on one node doing all the work. Pretty much the worst-case scenario for Apache Spark.

When reading the data, you'll actually get a more rational number back and not the one you might expect. The system is helping you out by splitting the incoming data into reasonable chunks based on the default settings mentioned previously. You can't stuff an infinite amount of bytes into a single partition.

You might think that it's a good idea to go crazy in the other direction. No, it's not. The reason is you get a lot of overhead in handling all the tasks. Let's see what it looks like in our example:

```
df = spark
        .read
        .option('delimiter',',')
        .option('header','False')
        .csv('/databricks-datasets/airlines/part-0001*')

df.repartition(10000)
        .write
        .mode('overwrite')
        .csv('/temp/csvforfun')

df = spark
        .read
```

```
        .option('delimiter',',')
        .option('header','False')
        .csv('/temp/csvforfun')
```

```
df.count()
```

You'll notice that it takes a while to go through all the files at writing. Counting to 10,000 takes a while even for a computer, especially when the counting is distributed. When reading Databricks will split the job into 200 tasks, which is a default value.

The reason it takes time isn't that strange. If you slice work into too small pieces, you'll end up spending more time on scheduling the work than actually doing it. You need to find a good balance.

So you might wonder how many partitions are good. It depends on the data and your nodes. A rule of thumb is three to four tasks per core. That might not always be correct, but it's something to start with.

As mentioned, Databricks is pretty good at figuring this out by itself. There might be instances where you find you can help out. If performance is poor or statements even fail, it might be a good idea to look at the default settings and the number of partitions used. Ones you might want to look at specifically are

```
print(spark.conf.get("spark.sql.shuffle.partitions"))
print(spark.conf.get("spark.default.parallelism"))
```

Another thing you might want to do is to define the partition key and thereby explicitly tell Databricks how to split the data. This can make sense in many cases. You might for instance know that you only need a small part of the data and can then use this as a filter:

```
df = spark
        .read
        .option('header','True')
        .option('inferSchema','True')
        .csv('/databricks-datasets/airlines/part-00000')
```

```
df.write
        .partitionBy('Origin')
        .saveAsTable('originPartitioned')
```

```
%fs ls /user/hive/warehouse/originpartitioned/
```

```
df2 = spark.sql('select * from originpartitioned where Origin = "SAN"').
explain()
```

Here we tell Databricks to partition the data by the Origin column and then store it. When we look at the actual structure, we see that every single origin airport got its own folder.

The benefit of this is shown in the last command, where we look at the execution plan for a specific query. As we're asking for a specific value in the Origin column, we only need to read the related data. No need to care about all other files. This is what's giving it away:

```
PartitionFilters: [isnotnull(Origin#305), (Origin#305 = SAN)]
```

In some cases you might have multiple keys that you want to partition on. That's possible as well. You just need to add the extra columns to the partitionBy clause like this:

```
df.write
    .partitionBy('IsArrDelayed','IsDepDelayed')
    .saveAsTable('arrdepPartitioned')

%fs ls /user/hive/warehouse/arrdeppartitioned/
%fs ls /user/hive/warehouse/arrdeppartitioned/IsArrDelayed=NO/
```

As you see we now have subfolders beneath the first partition, containing the other one. You can do this sub-partitioning scheme in multiple layers. Don't overdo it though. Pick something that suits your actual needs in the next step of the process.

One word of caution: If you partition data by yourself, you increase the risk of introducing skewness. We've mentioned this early in this chapter, but it's worth bringing forth again as it can be a real showstopper. You don't want to have uneven partitions. If you do, you risk sending much more work to one core, creating a bottleneck.

Summary

This was another big one. We've gone through a lot of additional Apache Spark functionality in this chapter. With all these in your arsenal, you'll be ready to handle most of the extract, transform, and load jobs you'll come across.

After a bit of introduction, we took a brief look at the Spark UI, after which we did a deeper dive into cleaning and transforming. Among other things, we investigated pivoting, exploding, and how to handle null values.

We followed that up by investigating optimization tricks like compression and caching. After a warning about using too many functions, we learned how to do them without declaring them in advance.

Finally, we spent a bit of time on storing data. By using different save modes, we learned how to append and overwrite data and manage external data. Then we played around a bit with the partitioning options available in Apache Spark.

In the next chapter, we'll look at how to connect to other systems.

CHAPTER 9

Connecting to and from Databricks

We've already looked at getting data into Databricks via files. A large amount of data will probably be pulled or pushed directly from and to the sources. This can easily be done using ODBC and JDBC.

In this chapter, we'll look at how we set up these solutions on Windows and Mac. We'll look at a few examples with commonly used tools on the market to see how it works.

We'll then flip the table and use JDBC to pull data into Databricks from a few different database systems. In the end, you'll have a good understanding of how you can ship data directly both to and from Databricks.

Connecting to and from Databricks

As you've probably realized by now, you can do your analytics work in Databricks. Most data won't be created there however, as it's not well suited as a platform for an operational system.

So it's common to access Databricks from another tool. Maybe you want to play around with data in a spreadsheet or in a visual tool like Tableau; maybe you want to further process the information in another data engine. Whatever your need, the ability to get data to and from Databricks has to be there. And it is.

One way of handling data transportation is, as we've already discussed, by using files. It is easy to do, decouples the systems, and doesn't require a direct network access.

It is, however, not convenient. This is especially true if you don't exactly know what you're looking for and want to play around with the data. It's an additional step you have to go through every time. Also, for many tools, you would have to use the terrible CSV alternative.

For those reasons it is sometimes better to create a direct connection. This gives you access to all information you have stored in Spark instantaneously. You send a query, and Databricks processes it and then returns the information to your client.

Open Database Connectivity, ODBC, and Java Database Connectivity, JDBC, are the most common generic way of getting data out of any database. It's true for Databricks as well. You can use either of them to connect to and from the system.

Both work in pretty much the same way. They offer a programmatic interface for communicating with databases. For them to work, you need to install a client package, a driver (which is totally different from a Spark driver). You then use this driver as a middleman when connecting to an external system. The driver contains everything needed to communicate with the target database.

The same solution works in the other direction as well. It's not just for other systems accessing Databricks; you can use it to pull data as well. While files, streaming services, or ETL tools might be more common for transporting data, you can use ODBC and JDBC in this direction as well.

Note that a direct channel to the clusters isn't always preferred. You can sometimes get better performance with files, if you can use a good file type. Another issue is that most client tools can't handle the startup time you get if the cluster isn't up and running. They'll just return an error code. And since the cluster has to be up, it might turn out expensive over time.

For deployed integrations that should work over time, you should consider tools made for that. Sqoop, NiFi, and Kafka are three popular, open source programs that work well with Databricks.

Getting ODBC and JDBC up and running

To get ODBC and JDBC running with Databricks, you need to prepare a number for things. First of all, you need to get the correct driver. There are a few that would probably work, but the one you want for connecting to Databricks is Simba's version. It's the best. Also, Databricks recommends it.

To get access to the drivers, go to `https://databricks.com/spark/odbc-driver-download`. Hand over your contact information. Fill them in and submit them. You'll get a Thank You page and after a little bit an email in your inbox. In it you'll get links to all versions of the drivers. ODBC drivers are target platform specific, while JDBC is the same across environments.

It's worth mentioning that the Simba Spark drivers are not free of charge normally. Databricks has made a deal with Simba to offer them to you without any additional cost, but note that you are only allowed to use them with Databricks. If you have your own Spark cluster, it won't be covered by the license you agree to here.

Creating a token

The second thing you need to access Databricks from your client using ODBC or JDBC is a token. This is a secret key that you use instead of a login and password in many tools and will identify you and the workspace you want to work with. So let's start by creating one of those.

In the top-right corner at any Databricks screen, you'll see an account button in the shape of an upper body. Click it and choose User Settings. The first tab of the screen you come to is Access Tokens.

Click the Generate New Token button. This will bring up a pop-up window with two questions. The Comment field is for you to use as a reminder. Put a few words there so you know why you created this token. I usually type in ODBC or JDBC plus the tool or system that I will use it with. That way I know what to remove once I stop using the tool.

Next up is the Lifetime. This is more interesting. While you can just decide to type in nothing and have the token work forever, it's a good idea to limit the life span. It'll force you to revisit the token from time to time and evaluate whether it's still needed or not. The default choice of 90 days is not a bad one. You'll be able to see the expiration date on the overview screen.

Pick a length and click Generate to create the token. You'll get it in a field and already marked, ready for copying. The reason is that this is your only chance to actually get the key. Once you click Done, it's gone and you'll never be able to get it again. Of course, you can get a new key if you lose one.

Anyways, we need a key so copy this one and keep track of it. We'll come back to it multiple times in this chapter so don't lose it. If you do, just create a new one. If so, click the small x to the far right side on the row of the token you no longer need. You want to close unnecessary openings whenever possible.

Preparing the cluster

Finally, no matter what client you plan on using, you will need to set up a cluster to handle incoming requests. You don't need to do anything special to it; just create one like we've done in the past. I personally like to have a separate cluster for ODBC/JDBC requests, but you don't have to do that.

Once it's running, expand the Advanced Options at the bottom of the cluster page. This will give you a few more options, including a row with links. One of those links is called JDBC/ODBC. Click it. This is where you'll find the base information needed to create connections.

One drawback with doing this on Databricks is that the cluster has to be up and running. If you try to access the data via ODBC/JDBC with the cluster down, it'll start up. In most cases the client tool will give up before that startup process is done, however.

One thing worth noting is that a call to the cluster will start it up. So while you might get a connection error if it's down in the client, Databricks will start it up. Don't forget to shut it down so you don't generate unnecessary costs.

Let's create a test table

We don't really need a specific test table, but let's create one anyways. That will make sure that all the examples will just work. So open a new SQL notebook so that we can build a small object for our ODBC adventures:

```
%sql
use default;
create table if not exists odbc_test (
headline string,
message string);
```

```
insert into odbc_test values ('ODBC test...', 'worked!');
select * from odbc_test;
```

Nothing fancy here. We just create – if it doesn't already exist – an empty table with two columns in the default database. Next up, we populate the table with one row. Like always we validate to make sure it actually worked. You never know if Spark will one day trick us.

Now you have everything you need to connect from your client tools fixed on the Databricks side. What remains is to get the drivers running on your local Windows or OS X machine.

Setting up ODBC on Windows

Installing ODBC on Windows is easy, thanks to the built-in support for it. Start by downloading the driver and then double-click one of the ODBC files. Make sure you pick the right one. Just because you have a 64-bit operating system doesn't follow that you should use a 64-bit driver. It's the client software that dictates the version. Luckily you can have both installed at the same time.

Whatever version you choose, you'll be guided through a normal installation process. Just click through it as there is nothing you need to change. After a few clicks and minutes, you'll have Simba ODBC installed.

Next up, start the ODBC Administrator client on your machine. Click the Add button to the right and select the Simba Spark ODBC Driver. Click Finish to get to the configuration view. While it might look complicated, it really isn't. You just have to fill out a few of the fields for Databricks.

Name is important as this is what you'll reference in your client tools. Description is up to you, of course. It doesn't matter what you write, but try to actually put something helpful in. The Spark Server Type should be SparkThriftServer. In my case it states this is for Spark 1.1 and later.

Now you'll need the cluster configuration view ready in the Databricks user interface. Several of the fields in the ODBC configuration will be filled in with data from the Databricks cluster view. Keep it open in the background as you copy and paste the information into the ODBC configuration window.

The Host is the Server Hostname you're using. Same goes for Port, which should be 443. Just copy them straight over to the ODBC setup. Database dictates which the default database should be. You can leave it as is if you don't have any specific target database in mind.

The Authentication Mechanism should be User Name and Password. This will gray out several fields. In the User Name field, type in the word "token". As you might guess, the Password is the token you previously created. Don't forget to check the Save Password option unless you want to type in the token every time.

Now there is just one more thing. Copy the HTTP Path from Databricks and click the HTTP Options in the ODBC configuration view. Paste the HTTP Path into the field with the same name. Click Ok and then Test to verify that everything works. Remember that the cluster has to be up and running for this to work. Save and you're done with the preparations.

Setting up ODBC on OS X

Handling ODBC on OS X is not as easy as on Windows. The reason is that there is no built-in software to support you. So you'll need to dig around in text files manually. Luckily it's not *that* complicated.

Download the driver file, open it, and execute the package inside. You'll go through a normal process with a license and multiple steps to get the software onto your computer. There is nothing you need to change, so just click through it.

While you can store the configuration files in multiple places, the easiest is to use the default /etc folder. That way you don't need to change any environment variables, and it'll be easier for others to use the same setup. The drawback is that you'll change things system-wide. Also, you'll need to do the changes as an admin user. If you want to store the files in another place, use the environment variables ODBCINI and ODBCSYSINI.

The first file we care about is /etc/odbcinst.ini. You'll notice that it's already prepopulated with two sections by the installation: ODBC Drivers and Simba Spark ODBC Driver. It should look like the following. If there is nothing in there, you can add this yourself:

```
[ODBC Drivers]
Simba Spark ODBC Driver = Installed

[Simba Spark ODBC Driver]
Driver = /Library/simba/spark/lib/libsparkodbc_sbu.dylib
```

The first section, ODBC Drivers, tells the system that the actual driver is installed for the driver name specified in the next section. You don't need to use the driver name specified though. If you prefer, you can shorten them or just add a new one using the same driver, like this:

```
[ODBC Drivers]
Simba Spark ODBC Driver = Installed
Simba = Installed
```

```
[Simba Spark ODBC Driver]
Driver = /Library/simba/spark/lib/libsparkodbc_sbu.dylib

[Simba]
Driver = /Library/simba/spark/lib/libsparkodbc_sbu.dylib
```

Next up you need to add information to the /etc/odbc.ini. This is where you define all the parameters needed to connect to Databricks. You should have the ODBC/JDBC tab on the cluster detail page open when doing this. There's a lot of information there you'll need:

```
[Databricks]
Driver=Simba
Server=<Server Hostname>
HOST=<Server Hostname>
PORT=443
SparkServerType=3
Schema=default
ThriftTransport=2
SSL=1
AuthMech=3
UID=token
PWD=<Your token>
HTTPPath=<HTTP Path>
```

The first line here is the data source name (DSN) you want to reference later. It can be whatever you want. For Driver you point to the section names we created in odbcinst. ini. So while we use the shortname Simba here, you could also use Simba Spark ODBC Driver.

Next up you need to define the hostname, which you'll find under the JDBC/ODBC tab. Schema is the default database you want to connect to; PWD is the token you created earlier in the chapter. Those are the only things you should change. Leave the rest alone. Those parameters just define how the connection to Databricks should be made.

As everything is done in text files, there is no test button, like on Windows. While you could download an ODBC manager, it's far too cumbersome. Instead, let's write a small Python script on your client machine:

```
import pyodbc
con = pyodbc.connect('DSN=Databricks', autocommit=True)
cur = con.cursor()
cur.execute('select * from odbc_test')
for row in cur.fetchall():
 print(row)
```

The pyodbc library gives us the capability we need, so we import it. Next, we create a connection by referencing the section name we created in odbc.ini. The autocommit part has to be there as we can't handle transactions. Then we create a cursor and attach a SQL query. Finally, we loop through the result.

If you get a "Data source name not found" error, it's probably due to an environment variable issue. Download and install odbcinst (yes, it's a small utility with the same name as the file) and run the following command to see where your computer is looking for the odbc.ini and odbcinst.ini:

```
odbcinst -j
```

As mentioned earlier, you can control where OS X should look by using the environment variables ODBCINI and ODBCSYSINI. They are sometimes pointing to a different location on systems with a lot of driver software installed (which is why I don't recommend changing them – it leads to confusion). They should look like this. Note that one is pointing to a folder and the other to the file:

```
export ODBCINI=/etc/odbc.ini
export ODBCSYSINI=/etc
```

By the way, if you don't want to go through the hassle with the ini files, there is a trick. You can actually type out all the parameters directly in the connection call. If you just want to connect to a cluster once, this might actually be easier to do:

```
import pyodbc
con = pyodbc.connect('DRIVER=/Library/simba/spark/lib/libsparkodbc_
sbu.dylib;Host=<Server Hostname>;PORT=443;HTTPPath=<HTTP
```

```
Path>;UID=token;PWD=<Your token>;AuthMech=3;SSL=1;ThriftTransport=2;SparkSe
rverType=3;', autocommit=True)
cur = con.cursor()
cur.execute('select * from odbc_test')
for row in cur.fetchone():
 print(row)
```

This is pretty much identical to what we did earlier and will work with no odbc.ini or odbcinst.ini. The only real change is that we do a fetchone instead of a fetchall, which is just to let you know both exist. Don't forget though that if you change anything in the connection string, it's much easier to update one odbc.ini file than hundreds of scripts. So don't use this method for stuff you'll use frequently.

Connecting tools to Databricks

Microsoft Excel on Windows

Microsoft Excel is probably the most commonly used tool in the analytics area. It doesn't really matter how good your dashboard is or how smart your machine learning implementation is – the data seems to always end up in Excel one way or another.

Start the tool, and open a new spreadsheet. In the Data tab, click the From Other Sources button and choose the From Microsoft Query option. This will open a new window in the Microsoft Query tool. Don't worry if it looks strange. If you weren't around in the 1990s, this is what tools used to look like back then.

In the middle of the window, you'll have a list of data sources from the ODBC manager. If you did everything right in the setup explained earlier, you should have your Databricks connection available. Select it and click Ok. Like always, make sure your cluster is running.

After a minute or so, you'll get a list of all the tables available in your default database. If you want to see objects in another database, click the Options button and change to the preferred database in the Schema dropdown menu.

Once you've found the table you want, select it and click the rightward-facing arrow followed by Next. Create a filter if you don't want all of the data and continue with Next. If you want the data ordered before delivery, let the Query Wizard know. Then click Next again and Finish.

After a few moments, you'll finally be back in Excel. Decide where you want the information to go. Once you've done that, Databricks will start sending data to Excel, and after a while you'll have everything in your spreadsheet.

As with all external sources, you can refresh the material from within Excel. The button is called Refresh All and can be found under the Data tab. Clicking it will resend the query to Databricks and update the information.

Microsoft Power BI Desktop on Windows

While Excel might be the traditional tool for many hard-core analysts, graphical tools like Qlik and Looker are getting more and more popular. They make it possible for more people to play around with data in an easy-to-use way.

One of the more interesting tools in the segment is Microsoft's Power BI – especially if you're running Databricks on Azure, of course. The reason is it offers a feature called DirectQuery. Instead of pulling all the data to the client before processing, you can have Databricks do the heavy loading and just display the results. This is very helpful if you want to play around with billions of rows in a visual way without code.

It's also available to download free of charge, at least the Power BI Desktop version. On top of that, it won't cost you more than a handful of dollars or euros per month to get the Pro version. To get it, look for Power BI on Microsoft's web pages.

Start the Power BI Desktop and click the Get Data button. In the search box, type in spark and you'll get three options. Select the one that is just named Spark. Then click Connect to get to the configuration stage.

First up is the server name. For whatever reason Databricks doesn't provide you with the link used here so you have to construct it. While it's not hard, it's annoying as another box in the user interface would have solved it. The link should look like this:

```
https://<Server Hostname>:<Port>/HTTP Path
```

So, for instance, a complete URL would look like this:

```
https://westeurope.azuredatabricks.net:443/sql/protocolv1/
o/111111111111111/1111-111111-hosts111
```

Type it in and select HTTP as the protocol. In the Data Connectivity mode, use DirectQuery. Note that if the table you select is very small, Power BI will suggest that you download it instead. This is good advice. Under normal circumstances, you don't want to bounce small queries back and forth. Use DirectQuery for the heavy loading.

The user name is the literal "token", and the password is the personal token you created earlier. Enter the information into the fields and click Connect. This will open the Navigator window where you'll see all the tables available to the cluster.

Check the box to the left of the table name to select it. Then click Load to get it into Power BI. No matter how big the table, this will be pretty fast. Not blazingly so, but still better than you'd expect for a large table. The reason, as mentioned, is that you didn't really get the data but only the definitions.

Once the fields are in the tool, you can use them like any other data. I recommend reading a huge table and creating a bar chart plus a slicer. Using those two tools, you can see how changing a filter will execute a query in Databricks and update the graph. This enables analysts to actually work with the lowest grain of data if they need to – or want to, whichever is more frequent.

Tableau on OS X

The biggest visualization tool on the market is Tableau. It's used by businesses everywhere and offers a wide range of features for data analysts. While there is a free version, it unfortunately doesn't allow connections to very many sources. So you'll need a full version or a trial to run this test.

Start the program, click the Data menu option, and add a New Data Source. This will give you a long list of options. It's possible to use the JDBC and ODBC connections, but there's also a direct link. Search for Databricks and select it when it pops up.

Use the Server Hostname and HTTP Path from the JDBC/ODBC tab. The user name is the word "token", and the password is the token key you created earlier. With everything filled in, make sure the cluster is running and click Sign In.

Tableau will process your request and return an empty workspace. Open the Schema dropdown list and click the small magnifying glass. This will populate all the databases. Select one and you'll get a chance to select tables. Do the same operation here, and you'll see all tables in a list.

Pick a table and drag it into the large field up and to the right. It'll populate the lower-right part with columns. Click the Update Now button to get the actual data into the tool. Then click Sheet 1 in the lower-left corner, and you'll get to the main work area.

You'll see the dimensions and measures on the left-hand side and can start building reports on the right-hand side. All the data is imported and there is no live connection. To refresh data, go to the Data menu, select the correct connection, and click Refresh.

CHAPTER 9   CONNECTING TO AND FROM DATABRICKS

PyCharm (and more) via Databricks Connect

The notebook developer environment is one of Databricks's strong points. It's not perfect in all situations though. It's, for instance, not very well suited for advanced coding. If you do a lot of that, other tools can be more helpful.

Luckily you can connect a traditional development tool to Databricks instead. One popular choice for Python is PyCharm. It's a nice editor with all the features you would expect from a normal programming environment. You can get the community edition for free from jetbrains.com.

To access Databricks from PyCharm, you need to use something called Databricks Connect. It pretty much works like any other library. You import it, use it to access Spark, and run code on it as if you were working in a notebook.

Like always you need to have a cluster ready. But there is one wrinkle. You need to match both the major and minor version numbers with the client. While I can tell you it's Python 3.7 in Databricks Runtime 6.x, it's always best to check. In a Python notebook, write

```
import sys
print(sys.version)
```

In the same way, you need to get the version number on your client. So type in the same command in your local Python installation to make sure that the major number and the first number to the right of the decimal point match.

Of course you can create an environment with a specific number on your client with virtualenv or conda. Those tools make it possible for you to run multiple separate Python setups with different versions on the same machine. I personally use Anaconda on my laptops.

To install the Databricks Connect library, you need to get rid of Pyspark as Databricks will use its own. So let's use the pip installer to remove it from your client system. Note that you might not have it, in which case pip will return a warning. From the command prompt or terminal, type in

```
pip uninstall pyspark
```

Next up, we install Databricks Connect. The version number in this command should be the cluster version number, not the Python one. I'm for instance working with a 6.2 cluster when I'm writing this:

```
pip install -U databricks-connect==6.2.*
```

CHAPTER 9   CONNECTING TO AND FROM DATABRICKS

It'll take a little bit to install, but after a minute or so you're ready to continue. The next step is to configure the setup so that the library knows what cluster to connect to. Like before it's good to have the cluster information at hand:

`databricks-connect configure`

Start with reading through the license and decide if you can accept it. If you don't, you won't be able to continue. I wish I could say something about it, but I'm not a big fan of click agreements and must admit I might not have read all of it…

If you accept, it will in order ask you for the host, the token, the cluster ID, the organization ID, and the port. As there are a few new ones, let's quickly go through them. Don't worry about making a mistake. You can redo this multiple times.

The host is https:// and the region you're in. Just look at your address bar in the web browser. It should be something like westeurope.azuredatabricks.net. You can also find the value under the JDBC/ODBC settings.

Token is the same token we've used earlier. Just paste in the long string of random characters. As mentioned before, you can just create a new one in case you've lost your key or want a new one for this use.

Every cluster also has an ID. If you look under Advanced Options and the tab called Tags, you'll find it in the form of ClusterID. It'll look like 1111-123456-word123. This is what Databricks Connect is asking for. You can also see it in the URL if you're at the cluster configuration page.

The Org ID is short for organization ID. If you look in your address bar in the browser, you'll notice there's a string just after the domain starting with o=. It's a long line of numbers, like 1111111111111111. This is the Org ID and it's only relevant for Azure users and not needed for AWS.

Finally there's the port. It's 15001. Just type all this information in, and you'll get a few confirmation rows telling you where the configuration file is and a few suggested tools. Copy the last suggestion and run it:

`databricks-connect test`

This will make sure that you can connect to the cluster and run a job. Don't worry if you get a number of warnings though. The only thing you should care about right now is that you get an "All tests passed" and a few SUCCESS messages in the end.

Next up it's time to test this in PyCharm. Start the tool, open the Run menu, and select Edit configurations. Expand Templates, select Python, and click the icon to the right of the Environment variables text field. Click the plus sign and add

PYSPARK_PYTHON=python3

This is to make sure the right version of Python is run and might not be necessary in your environment. Also, you need to make sure that the project interpreter is the one you've been working with. If you're not sure how to do this, here are the steps.

Click File and then Settings. Expand the Project structure. Select the Project Interpreter row and on the upper right-hand side click the cog. Select Add. Click the System Interpreter line and then Ok. This will make sure your project runs on the base version of Python:

```
from pyspark.sql import SparkSession
spark = SparkSession\
.builder\
.getOrCreate()

print("Starting")
df = spark.sql('select * from odbc_test')
print(df.count())
print("Finished")
```

Run this code, and you'll get a count of the number of rows in the table. As you can see there are a few things that are different here compared to normal Databricks notebooks. The main thing is that we need to create the Spark session ourselves.

The important thing is that the jobs are run on the server. That means that you don't have to transport data to your client or use its limited resources. You just get the results. It's a nice way to code in a proper development tool and also have access to processing power.

By the way, if you get warnings but the code works, you can just ignore them. They'll probably tell you that your installation is missing something that isn't critical for what you want to do.

If you want to get even closer to Databricks in your local environment, you can also install the dbutils toolkit. This will give you access to the file system, passwords, and other features which can be useful if you want to create real solutions. Pip comes to the rescue as usual:

```
pip install six
```

Most likely you'll already have it installed as it comes with Databricks Connect. To test it, let's look at the file system. As you might remember, we can run some traditional unix commands to list files:

```
from pyspark.sql import SparkSession
from pyspark.dbutils import DBUtils
spark = SparkSession\
.builder\
.getOrCreate()

dbutils = DBUtils(spark.sparkContext)
print(dbutils.fs.ls("dbfs:/"))
```

This will return the content of the DBFS root. As you can see, we still need to create the Spark session and also a context explicitly. Once that's done, we can just run the command as if we were doing it in a notebook.

There is one more thing worth mentioning. This same method works for Jupyter, Eclipse, Visual Code, and several other tools as well. It's also available for use with Scala and Java. So PyCharm is just an example and not the only choice. You can probably continue to use the tool you prefer.

Using RStudio Server

Although I'm not at big R user myself, I recognize it's a popular tool and that a lot of people like it. Most of all, they like a development tool called RStudio. While you can connect to RStudio using Databricks Connect, like with PyCharm, it's even better with RStudio Server installed on the cluster. Let's look at how to do that.

First of all, you need to create an initialization script. You can add this in multiple ways, just like with data. The easiest way is to write the code from within Databricks. While there is no direct editor, you can do it indirectly. In any Python notebook, type in and run this, as suggested by Databricks:

```
%python
script = """#!/bin/bash

set -euxo pipefail
```

```
if [[ $DB_IS_DRIVER = "TRUE" ]]; then
  apt-get update
  apt-get install -y gdebi-core
  cd /tmp
  wget https://download2.rstudio.org/server/trusty/amd64/rstudio-server-
  1.2.5033-amd64.deb
  sudo gdebi -n rstudio-server-1.2.5033-amd64.deb
  rstudio-server restart || true
fi
"""

dbutils.fs.mkdirs("/databricks/rstudio")
dbutils.fs.put("/databricks/rstudio/rstudio-install.sh", script, True)
```

If you go to https://rstudio.com/products/rstudio/download-server/debian-ubuntu/, you'll find both the commands and the latest version of RStudio Server. That said, older versions are available as well so the preceding example will work for years to come.

Whatever version, this will create a file called rstudio-install.sh in the /databricks/rstudio folder. The actual script is the text between the apostrophes. The rest is just commands to create the text and save it into place.

What the actual script does is an installation of RStudio Server 1.2 on the driver. The DB_IS_DRIVER flag keeps the worker nodes away. In the end there is a startup of the actual software. All this is only a script though. Running the preceding code won't actually execute it. That you do in the cluster.

Create a cluster you intend to use for this RStudio Server setup. Make sure you have the automatic termination turned off. For whatever reason, you can't have this feature enabled with the RStudio Server setup.

Go into the detail page for the cluster and click the Edit button. With the fields unlocked, open the Advanced Options and click the tab Init Scripts. Here you can add code that you want executed at cluster startup. Type in dbfs:/databricks/rstudio/rstudio-install.sh and click Add. Then confirm and start the cluster.

What will happen now is that once the cluster is up and running, the preceding script will be executed and RStudio Server installed on the driver node. Next up, you need to do a setup. Note that I'll assume you're running the open source version of RStudio.

In the cluster details, click the Apps tab. You'll have a button for setting up RStudio. Click it to get started. Click the show link to access the password. Copy it. Then click the Open RStudio UI link to open up RStudio. Log in.

That's it! From here you can install the SparkR library and run Spark on Databricks. You'll have access to all the underlying nodes. Of course you can use Sparklyr as well, just like most of you R folks out there like it. Here's a short R example just so you can verify that everything works:

```
require(SparkR)
sparkR.session()
df <- sql("select * from odbc_test")
head(df)
count(df)
```

This will create a Spark session, read the data from my_tab to the df dataframe, and run two commands on it. The important thing here is that everything you run is executed on the driver and, if necessary, on the worker nodes.

Accessing external systems

The preceding examples have all been about connecting to Databricks from other systems. Now let's flip that around and look at how we can access data from other systems using Databricks instead.

We've already looked at this a bit when we connected to the cloud data storage. This is a common scenario, but in many cases you want to access relational databases directly. While unstructured data is all the rage nowadays, most useful data is still in traditional systems.

Most systems will require you to use some version of JDBC. While you sometimes can use other methods, this is a way that almost always works. So if you know the basics of how to do it, you'll be ready for a new system when it arrives.

Remember that while a direct connection is nice and easy, it's not always a good thing. You connect the two systems together in a way that will make it hard to separate them later – easy if it's one job and a nightmare if there are hundreds.

Another small detail: For connections like these to work, you need to have access to the databases you want to collect data from. This might mean you need to set up virtual networks and VPN, which is outside the scope of the book. The examples here assume that the connection part is already handled.

CHAPTER 9   CONNECTING TO AND FROM DATABRICKS

A quick recap of libraries

In an earlier chapter, I briefly mentioned that you can work with libraries in Databricks. When you want to pull data from an external source using JDBC, you'll use this a lot. So let's do a quick recap of what it is and how it works.

Libraries are the easiest way to expand the feature set of Databricks. You can pull in functionality from huge repositories online, like PyPI, Maven, and CRAN. On top of that, you can bring your own files. Jar files, Python Egg files, and Python Whl files are all welcome.

To install a new one, click the workspace button in the left-hand bar in Databricks. Move to the place where you want to store the libraries. A subfolder in the Shared folder is a good place, for instance.

Wherever you want to store it, select the folder and click the downward-pointing arrow to the right of the folder name. Click Create and then Library. This is your main work area for adding libraries.

On the top you select a source. You can either pick the file yourself and upload it or link to the library you want on one of the big repositories. If you want to look at what's available, you can go to `https://pypi.org/`, `https://mvnrepository.com/`, and `https://cran.r-project.org/`.

Once you've installed the library of your choice, you'll have to attach it to the clusters you want to have access to it. This is crucial. The libraries are attached per cluster, not workspace.

If you only want to use a library once, you don't have to go through this process. There are a couple of shortcuts. Unfortunately they only exist for PyPI and are manually uploaded ones:

```
dbutils.library.installPyPI('datarobot', version='2.19.0')
dbutils.library.restartPython()
```

The first command installs any library you want directly from the notebook. In this case it's a library for working with DataRobot. The second command restarts Python, which will clear the state of the notebook. It is, however, frequently needed for the installed library to work.

Connecting to external systems
Azure SQL

If you are using the Azure cloud solution, you're probably familiar with Microsoft's Azure SQL. It's a modern, cloud-based database system with close ties to the traditional SQL Server product that has been around for more than 30 years.

While Databricks is great at processing large amounts of data, it's not built to be used by operational systems. It also lacks many features that you'd expect out of a traditional relational database management system. So something like Azure SQL makes sense as a complementary product.

Databricks and Azure SQL work well together. Connecting to the remote system is easy as the drivers are installed by default in Databricks. You can just start coding directly. So let's try it out. You need an Azure SQL database for this example to work, of course:

```
hostname = "<yourhost>.database.windows.net"
database = "<database>"
port = 1433

url = "jdbc:sqlserver://{0}:{1};database={2}".format(hostname, port, database)
prop = {
  "user" : "<username>",
  "password" : "<password>",
  "driver" : "com.microsoft.sqlserver.jdbc.SQLServerDriver"
}
sql = "(select * from sys.tables) a"
df = spark \
.read \
.jdbc(url=url, table=sql, properties=prop)

display(df)
```

We start by defining the URL with the host, the port, and the database. Next up, we define a dictionary with our credentials and driver choice. The query is stored into a string, and finally we execute everything and return the result to a Databricks DataFrame which we show.

CHAPTER 9 CONNECTING TO AND FROM DATABRICKS

Oracle

In larger companies, Oracle still dominates on the database market. For that reason it's likely that you'll pull data from it directly or indirectly. If you have access to it, you can pull data to Databricks easily.

There are two ways to connect to Oracle. The first one is to use their cx_oracle library. That's what Oracle recommends for Python use and works well if you want to read smaller tables in a somewhat restricted way. Another option is to use their JDBC driver without the extra layer. Let's look at both options.

Let's start with JDBC. For this to work well, you should use Oracle's own driver that you can find at www.oracle.com/database/technologies/appdev/jdbc-downloads.html. You most likely need to use either the latest ojdbc8 or odbc10 driver. Note that the 8 does not indicate database version. Uncompress the file and locate the file you need.

Now you need to upload the file to Databricks. Go to your workspace, open your selected library folder, and click the downward-facing arrow. Select Create, followed by Library. Drag the ojdbc8.jar file (or whatever version you use) to the designated area and click Create.

In the next view, choose if you want it installed on a specific cluster or if you want it automatically installed on all clusters. Note that you only see the running clusters in the list. This is a bit strange but how it currently works.

If you want to do it the other way around, you can go to the detail page of a running cluster and find the Libraries tab. There you can both install and uninstall libraries you have available in the workspace.

Either way, once it's installed on the cluster, you can use the driver freely. The following query will pull the data from my_tab into the df DataFrame. The URL option explains how to access Oracle. The dbTable if for referencing tables or writing SQL. The user and password is exactly what it says.

The driver is where we reference the actual driver. You can try others if you want, but this one is pretty good. Finally, the fetchsize tells Oracle that we want to fill our blocks properly. The default is way too low and will make getting data slow:

```
df = spark \
.read \
.format("jdbc") \
.option("url", "jdbc:oracle:thin:@//<servername>:1521/<servicename>") \
.option("dbTable", "(SELECT * FROM dual)") \
```

```
.option("user", "<username>") \
.option("password", "<password>") \
.option("driver", "oracle.jdbc.driver.OracleDriver") \
.option("fetchsize", 2000) \
.load()
```

This is a rather simple example. You can do quite a few things to optimize loading of data using this method. The best part is that you get a Spark DataFrame immediately, in one go. So you could, for instance, load the data into DBFS with a write command.

As you might have understood, I much prefer this solution to the cs_oracle one. As I'll show in the advanced section of the book, this is an excellent way to load a large amount of data in an efficient way, if Oracle is the source, that is.

The other way to load data is cx_oracle. This library is maintained by Oracle, has a lot of features, and is what they'll point to if you want to connect to the database using Python. It's not preinstalled in Databricks though so you need to get it, which is easy. Installing it is a bit more complicated though as it requires a bit of wrangling.

Go the Create Library view, but this time you select the PyPI tab. Type in cx-oracle and click Create. Databricks will automatically pull the necessary files and install them. Like with JDBC, you need to select which clusters you want the library installed on.

Unfortunately this is not all. For cx_oracle to work, you need their instant client installed. So we need to create a little initialization script that helps us do all the work at cluster startup:

```
%python
script = """#!/bin/bash

if [[ $DB_IS_DRIVER = "TRUE" ]]; then
  apt-get update
  apt-get install -y alien nano libaio1 libaio-dev

  cd /tmp
  wget https://download.oracle.com/otn_software/linux/instantclient/195000/oracle-instantclient19.5-basiclite-19.5.0.0.0-1.x86_64.rpm

  alien -i oracle-instantclient19.5-basiclite-19.5.0.0.0-1.x86_64.rpm

  echo "/usr/lib/oracle/19.5/client64/lib" > /etc/ld.so.conf.d/oracle-instantclient.conf
  ldconfig
```

CHAPTER 9　CONNECTING TO AND FROM DATABRICKS

```
fi
"""
```

dbutils.fs.put("/databricks/oracle/oracle.sh", script, True)

This script installs all the dependencies for the Oracle instant client using the apt-get commands. It then downloads the software and installs it. Finally, we do a few configuration steps to let Oracle know where the libraries are located. The DB_IS_DRIVER makes sure we only install this on the driver node.

Before you use this straight up, make sure you're using the latest version. It can be found at www.oracle.com/database/technologies/instant-client/linux-x86-64-downloads.html. You'll also find some of the installation information there.

Once everything is installed, you can run a simple test. Let's see how many objects our user can access. It's a simple query that will verify that you can connect to your Oracle database using cx_oracle:

```
import cx_Oracle

tns = cx_Oracle.makedsn('<servername>', '1521', service_name='<service name>')
con = cx_Oracle.connect(user='<username>', password='<password>', dsn=tns)

cur = con.cursor()
cur.execute('select count(*) from user_objects')
result = cur.fetchone()
print(result)
con.close()
```

Nothing fancy going on here. We import the library and create a tns entry which we then use to create a connection. We open a cursor, execute a query, and return the result to a variable that we print. It works well, but the JDBC method is both easier to install and faster.

MongoDB

Let's try a NoSQL database as well. One of the most popular document databases around is MongoDB. Unlike relational databases, data is stored in the JSON format. It's convenient in some specific use cases.

MongoDB can be installed on-premise, but let's try the cloud-based Atlas instead as that's where a lot of the current action is happening. It's also easier to access in many cases as it is pretty much built for it.

Ok, so let's look at what you need to do. To start with you need to install two libraries: pymongo and dnspython. The latter isn't needed if you run your own MongoDB instance, only if you connect to MongoDB Atlas.

As explained earlier, just go to the workspace and create the libraries. Both are available at PyPI and can be added directly from the UI. Once that's done, restart the cluster. This is needed for the dnspython library to install properly:

```
import pymongo
import dns

client = pymongo.MongoClient("mongodb+srv://<name>:<pass>@<your-cluster>.azure.mongodb.net")
db = client.sample_mflix

cols = db.list_collection_names()
for col in cols:
    print(col)
```

Once those two packages are ready, you can start coding. This is a small example of how you can access the MongoDB cluster, connect to a database, and loop through its collections. As you can see it's pretty easy to get going once the libraries are installed.

Summary

Whoa, this was a somewhat dense chapter. While it's usually more fun to focus on the actual tool, in the world of data, integrations are as important. You can't get anything done without it.

In this chapter, we looked at how we can connect to Databricks using different client tools. Clusters are built to handle calls from programs like Excel, Tableau, and Power BI; and we tried them out.

We also looked at how to work with Databricks without using the built-in notebooks. By installing a simple tool, we connected PyCharm and opened up a new range of possibilities.

Then it was time to look at how we can connect to other tools from Databricks. Instead of using files or ETL tools, we can access data directly using JDBC. We looked at how that works with some common databases, like Oracle and MongoDB.

CHAPTER 10

Running in Production

If what you developed is good and not a one-off, you'll want to put it into production and have it run on a schedule. Valuable results should be delivered on a regular basis, possibly to another piece of software that'll execute on the findings. This topic is huge, and we'll only scratch the surface here by focusing on technical parts in Databricks.

You will want to run your code autonomously when it's time for production, so we'll spend a bit of time looking at jobs. We'll then dive into the built-in solution for running notebooks on a defined schedule.

While most work in Databricks is done through the user interface, there are times when the command line and application programming interface are a better choice. Luckily they're easy to use, as you'll see.

Of course we'll also take a deeper look at security and what you can do to protect your data and keys. Closely related to that is the pricing which will go up with the necessary security features, so we'll revisit that topic.

General advice

Before we get into the nitty-gritty detail of running jobs in Databricks, let's go through a few points that are relevant no matter what tool you use. The overall rule is to think about maintainability first and foremost. While this might seem obvious, it's far too common to see exceptions to this simple idea.

You might think that I don't live as I learn based on the examples in this book. The reason I'm not adhering to my own rules is for the sake of brevity. There would be a lot of extra code and pages with limited additional instructional value. So do as I say, not as I do, at least not in this book.

Assume the worst

Things will go wrong. It doesn't really matter how well you write your code and how clean the data coming in is. Sooner or later some unexpected thing will occur, there will be an exception, and your code will fail.

At the very least, make sure your code can handle errors. Add asserts, try/except clauses, and other tests to safely exit the program with good information to the calling program and the operations team. The less investigation they need to do, the faster the program will be up and running again.

The next level up from this is to do data validation and sanity checks on the information coming in. A crashed program is bad, but delivering erroneous data is worse. Sometimes it can be totally devastating. There are examples within high-frequency trading that have caused huge losses to companies.

Write rerunnable code

This is an extension of what we just talked about. You don't want to change a notebook or run parts of it manually in case of an error. If the operations team has to go into your code and change variables or empty temp tables, you didn't do your job right.

This is not the same thing as not being dependent on an earlier step. Make sure you verify that all conditions are met before you run your code. Keep track of the state in the code to minimize the need of unnecessary cells that need rerunning.

From a Databricks perspective, I have a small tip, which we'll go into later in this chapter. Break your code into multiple notebooks and run them in a chain. As a sidenote, Databricks actually doesn't want you to go beyond 100 cells in a notebook. I think you should split your code into parts way earlier.

Document in the code

You probably won't do this but you should. Every single developer, no matter if they code operating systems or do data science, tells themselves they'll document the code later. Most never get to it.

Here's the thing – your mind won't be in the same place next time you revisit the code. For me a weekend is enough to throw me off. I've also looked at old, undocumented code and rolled my eyes over how incoherent it is, just to realize it was actually me who wrote it in the first place.

Also remember that documentation is not about typing out the obvious. Just writing for the sake of writing is often worse than writing nothing at all as readers will have to parse more text with no added value. Compare the following two examples:

```
# This is a loop
for x in mylist:
 some logic
```

```
# This loop iterates over train stations, runs linear regression on the
  arrival times delays
# and tries to predict future arrival times. The output is stored into
  predlist.
# Note: I also tried logistic, lasso and ridge regressions but all
  performed worse.
# Example runtime: 18 seconds per loop on the PROD_SMALL cluster.
 for x in mylist:
 some logic
```

This second version is much better. You get a sense of what the loop is doing without actually having to look at the code. Ideally you should be able to pull out all the comments and understand the full flow of the notebook.

Write clear, simple code

Another step in the documentation process is to keep it simple. Once you learn a bunch of tricks, you might want to show them off in your code. This is great as long as it's necessary. In this case, you should document what's happening. If it's not necessary, try using easy code instead even if it's slightly less optimized. No one likes to dig through thousands of lines of nested SQL. Break it up.

Don't use generic variable and function names. It's far too hard to try understanding what it is that they contain without having to go through tons of code. So if your variable contains a list of stores, name it listOfStores or something similar, not just l as a short for list.

Keep it simple, in short. It's interesting to look at what really experienced coders produce. At a first glance, it looks like a beginner wrote it as they use the easiest, most expressive solutions. There's a reason for that.

Print relevant stuff

This one is for your sanity. Return a lot of information to either the screen or a log file. When things go wrong, or even if they don't, it's helpful to see what happened. Databricks does a lot of this automatically, but you should add more, especially if you have a lot of logic in a single cell.

Keep the logs. It's far too common to put a cleanup part at the end of the scripts. In that scenario you can't really investigate what the data looked like if something looks funny after the run. It's a much better idea to leave the temporary tables and output files. Instead start your script by cleaning them up.

Small tip: If you think there's too much output, make it conditional and return data based on an argument – something like a normal mode, a mode with warnings, and a debug setting for pumping out everything to the one looking.

Jobs

When you want to execute notebooks automatically in Databricks, you use a feature called Jobs. We've seen it mentioned earlier in the book, but never really got into it. Now is the time to look at it in more depth.

In essence this feature defines a template that you trigger when you want to run your code in the background. This tells Databricks what to run and how. The rest is being handled for you. You can also have it trigger on a given schedule for regular runs.

Let's create a job from scratch to see how it works. Start by clicking the Jobs button on the left-hand toolbar. This will take you to an overview where you'll see all the defined jobs. You can toggle between your own jobs and all jobs on the upper right-hand side.

Click the Create Job button. You'll get a template page where you can select a few things. Let's start by clicking the Notebook link to the right of Task. Pick one of your notebooks that is ready for running and then accept with Ok.

Next up, press the Edit link to the right on the Cluster information line. You'll end up on a cluster creation page where you can define the cluster as you normally do. There is one difference though, and that is in the Cluster Type section.

While you can choose to use one of your normal clusters to run your job, you will be better off cost-wise to create an automated cluster. This will make Databricks spin up a cluster of the size you specify, run your job, and then shut it down – perfect for our need. Confirm the preselected, small one and go back.

We can leave the scheduling for now. By not selecting anything here, the job will execute immediately and then never again. That suits our need for now. Finally we have the advanced options. We won't change anything here, but let's quickly go through them.

Alerts lets you send an email when the job starts and when it finishes. This might make sense if you want to get a warning when there's an error in your scheduled job. It's probably not what you want to use in a larger operation with many jobs running. In those cases external options calling jobs over the command line interface are usually better. More on that later.

The Concurrency option lets you limit the number of iterations of this job that can run at the same time. While the default of one makes sense in most cases, you sometimes want to run the same job multiple times with different inputs. If you have a generic data loading job, for instance, you typically don't want to just allow one load to run at the same time.

Timeout gives you an option to kill the job after a given amount of time. It's brutal as it'll just stop the job hard, but a good security measure. If you have something like an erroneous job with an infinite loop, it can run forever and cost you a fortune without this option. Don't be too conservative with it though. It's painful to see the script die unnecessarily.

Finally, the Retries option tells Databricks how many times it should try to run a job that failed and how long the waiting period should be. There are situations where the problem is temporary and a rerun helps. They are not too common though. If they're common, you should build in a safeguard for it in the script and not rely on this feature. Permissions we'll come back to later.

Leave all the options as is and click the Run Now link below the Active runs headline. The line will transform and show you a bit of metadata around the job you just triggered. You get access to logs, see the start time, and track the status. As you'll be waiting for the cluster to spin up, you'll probably see a pending state to start with.

Once the job is finished, the information is moved down to the list of completed jobs. You still have access to the information, and the status field now tells you if the notebook finished ok or not.

Click the name of the job to see the result of the run. The output is different from looking at a notebook. At the top you get metadata about the job, like start time, duration, and if it ran successfully. You also get information about the cluster, which can be helpful in case of failure.

The core part of the log is the output though. For each cell in the script, you see the code, the output if any, and the runtime. At the top there's a dropdown list where you can choose to see just the results – simple but very useful.

Scheduling

Running a job immediately and only once can be useful. It's when you schedule jobs that you get the real benefits of using this system, though. That's what you need to have solutions that run every day, week, or month. The solution for this in Databricks is aptly named scheduling, and we saw it when we created a job.

Underneath the surface, scheduling in Databricks is nothing but a user interface on top of good old Cron. If you are not familiar with Cron, it's a job scheduling system that's been around in the Unix world for a very long time.

One issue with Cron is that it is not built for running things once at a given time. There are other ways to do that, like the at command, but they're not built into Databricks. I hope this will be added eventually. For now you'll have to schedule one-time jobs externally. Or use dirty tricks, like removing the schedule from the script once it has run.

Anyways, let's define a schedule for the job we just created. Click it and click the Edit button for the schedule. This will bring up a window where you define how frequently you want to run the job.

Using the dropdown boxes, you can define when the job should run. The user interface here is not great. You should start by changing the second one before the first one to make this work well.

Luckily you can also look at the Cron syntax. If you know your way around it, you'll have a better time using it directly instead. Note that you'll still need to set the time zone though. For reference it's minute, hour, day of month, month, and day of week. The asterisk means "each."

To test the functionality, set the schedule to Every day at whatever time it is when you read this, plus five minutes. Don't forget to set the time zone correctly. The cron command for say 13:35 should be 35 13 * * * ?.

Now wait for the clock to turn. When the time comes, you'll see it in your lists, running. Don't forget to remove the schedule if you don't want it to run daily. In the main job list, you can order by schedule to see what's actively in the loop.

Running notebooks from notebooks

While you can put a lot of information into a notebook, it's not always smart to do so. If there's a lot of logic, it might be better to break it up into multiple parts. What you can do instead is to run the notebooks from an overarching notebook.

Let's try it. Start by creating a Python notebook containing a single cell with the following line of code. Let's call it Sub. As you can see, it will just return a fixed string back to the prompt. Now find this notebook in your Home folder:

```
print('Sub-notebook')
```

Right-click the file in the Home folder, or click the little arrow to the right of the name, to open a menu. In it, click the Copy File Path option. You'll now have the full path to the file you just created in the clipboard.

Next, create a new Python notebook named Main. This will be the calling one. In a new cell, type in the following command. Add the path as an argument. The timeout is also required to minimize the risk you'll end up in a costly hung state. Execute the command and see what happens:

```
dbutils.notebook.run(path = "<full path to your file>", timeout_seconds=100)
```

The output will be different from normal. Instead of getting the output you might expect, you just see Notebook job and a number. What happened is that the preceding command ran the other notebook and dumped the result into a log view, just like with the jobs.

Click the link to get to the log page. You'll see the metadata and the output from the one single cell in the other notebook. Now, wouldn't it be great if you could actually send that result back to the calling Main notebook? Good news! There is a way. Dbutils comes to the rescue again.

Open the Sub notebook again and replace the contents with the following lines. You can write both in the same cell if you like, but I prefer the return command to be in a separate cell. It doesn't matter from a functional perspective though:

```
returnvalue = 'Sub-notebook'
dbutils.notebook.exit(returnvalue)
```

Feel free to run the command. It should return Notebook exited: Sub-notebook. Now open the Main notebook again so we can capture the result. Change the content to the following lines of code. Then run it:

```
returnvalue = dbutils.notebook.run(path = '<full path to your file>', timeout_seconds=100)
print(returnvalue)
```

This time we collect the returning value into a variable and then print it. You still get the link to the log, but you also get an output row. Note that using the exit command will end the notebook. Cells below that point will not be run.

As you might imagine, using this setup, you can build logical chains for your jobs. You can return results to the calling notebook and decide what to do next. If you, for instance, want to load data and make sure it loaded ok, you could build something like this:

```
retSalesLoadNA = dbutils.notebook.run(path = 'load sales data NA',
timeout_seconds=100)
retSalesLoadEU = dbutils.notebook.run(path = 'load sales data EU',
timeout_seconds=100)
if (retSalesLoadNA == 'OK') and (retSalesLoadEU == 'OK'):
 retSalesProcess = dbutils.notebook.run(path = 'process sales data',
 timeout_seconds=100)
 if retSalesProcess == 'OK':
  retSalesValidate = dbutils.notebook.run(path = 'validate sales data',
  timeout_seconds=100)
   print(retSalesValidate)
  else:
   print('Data processing failed')
else:
 print('Data load failed')
```

And so on. This way you won't continue forward unless all previous steps finished correctly. By using this system of notebooks calling other notebooks, you can break your logic down into manageable chunks.

There is one thing in the preceding example that is bothersome though. The two data loading jobs seem pretty similar, and they probably are. It's unnecessary to have multiple notebooks differing only on the arguments. It would be much better if we could pass those arguments from the calling notebook. Enter widgets.

Widgets

Parameters or arguments enable you to write generic notebooks for a given feature, much like functions and procedures in normal programming languages. The way Databricks enables this is through a somewhat cumbersome functionality called widgets.

I write cumbersome, but they really aren't that complicated to use. It's just a strange way to combine two different use cases. Normally widgets in Databricks are used for selecting data within the notebook. We'll come back to this in Chapter 11. Here we'll just look at how to use the feature as a vehicle for passing arguments between notebooks.

Open the Main notebook and change the code to the following code chunk. As you can see we're just adding one thing, the arguments parameter. What we're sending is a dictionary object where we define the value of param1 to 1:

```
returnvalue = dbutils.notebook.run(path = '<full path to your file>',
arguments={"param1": 1}, timeout_seconds=100)
print(returnvalue)
```

Now open the Sub notebook so that we can handle the incoming argument. In the example we pick up the value for param1 using getArgument. Then we just pass it back as an exitcode for the calling notebook:

```
arg = dbutils.widgets.getArgument('param1')
dbutils.notebook.exit('You sent the value: {}'.format(arg))
```

Run the Main notebook, and you'll get the string you sent: the value, 1. If you change the number 1 in the call, you'll get a different number. Of course you don't have to use numbers. Strings work as well.

If you want to send over more complex objects, like lists, you need to do a bit of trickery. While it's usually better to rewrite the logic in the notebook, there are situations where this makes sense. Let's look at a somewhat dirty way of doing it. In the Main use this code:

```
plist = ['This','is','a','list']
returnvalue = dbutils.notebook.run(path = '<full path to your file>',
arguments={"paramlist": str(plist)}, timeout_seconds=100)
print(returnvalue)
```

Note that we do a conversion of the list using the str command. This will be sent over to the Sub notebook as a string. That means we need to revert it to a list on the other side. Open the notebook and use this code instead:

```
params = dbutils.widgets.get("paramlist")
wordlist = params.replace("'",'').strip('][').split(', ')
dbutils.notebook.exit('Your first word was: {}'.format(wordlist[0]))
```

209

We pick up the parameter just like before. In this case we use get instead of getArgument, but it's the same thing. I'm only showing it here so you can see both options are available for you.

Next up, we take the string and break it up into parts again using strip to remove the unnecessary characters and split to get the list. On the final line, we return the first word in the string.

If you don't see how the conversion from string to list is happening, you can run it step-by-step. Try running the following code in a separate cell. You'll get an output for each step of the process. The last command, split, returns a list:

```
l = str(['this','is','a','list'])
print(l.replace("'",''))
print(l.replace("'",'').strip(']['))
print(l.replace("'",'').strip('][').split(', '))
```

By the way, you can use an escape character in the replace command if you prefer not to mix apostrophe types. Instead of what I've used in my preceding example, you'd have to run the command like this:

```
wordlist = params.replace('\'','').strip('][').split(', ')
```

Now you hopefully have an understanding of how you can pass information back and forth between notebooks. While it is basic stuff, it gives you a lot of options for building large-scale solutions. Let's apply these things to the jobs.

Running jobs with parameters

Now that you have an understanding of how widgets work, it's time to run another job. This time we'll pass an argument to the notebook. As you might have guessed, we'll use the widget functionality. So let's schedule a notebook with just these lines in it. Call it Jobtest:

```
jobparam = dbutils.widgets.get("jobparam")
dbutils.notebook.exit('You passed the line: {}'.format(jobparam))
```

Create a new job and click the Select Notebook link. Pick Sub and then click Ok. This will give you an option of selecting parameters. Click the link. In the window that pops up, type in jobparam in the left box and Hello World in the right one.

Once you've typed in the two strings, click the Add button before Ok. If you don't, the parameter won't be saved. You should see the parameters spelled out beneath the jobname in the main screen if everything went fine. When you're ready, run the job.

Wait for the cluster to spin up, the job to run, and the status to come back as successful. If you don't want to just sit and reload the screen, feel free to add your email into the alerts settings box. You can also see the status on the Clusters page, under the Automated Clusters section.

Also remember that you can of course have a cluster running and use that for these jobs. Then you can test quicker, not having to start up a new set of nodes every time you run a job. In a real setting, don't use an interactive cluster, as explained earlier.

Once done you can click the name to get the result, just as we saw before. Filter on Results Only. In the output section, you should get something on the lines of Notebook exited: You passed the line: Hello World.

How about if you want to return a failure? Then you need to send back an exception. You can use a try/except to do it normally or create your own fail type. Create a notebook with the following code and run it as a job:

```
class myRunError(Exception):
  pass

raise myRunError('Forgot to handle parameters...')
```

Your job will come back with failure as a result. Of course you'll want to wrap this in some logic so you send back a proper exit if everything works well and an exception if it does not. Bonus point to you if you create good errors to make everyone understand what went wrong instead of a generic one.

In the same way you add parameters, you can add dependent libraries, like JDBC drivers. Just click the link and find the library you want to add to the cluster. Note that you need to have already downloaded them to your workspace.

This way you can do pretty advanced flows. You can call a notebook that in turn calls other notebooks, passing parameters back and forth. There is almost no end to the options you have here as you can do all your logic in Python (or Scala for that matter).

All this said, you probably shouldn't do too complex solutions within Databricks itself. While the functionality works well for small-scale stuff, it's a bit cumbersome to use and really not meant for advanced flows. It's hard to keep track what's happening.

So while you can build a lot of functionality into notebooks, you probably want to use something else for automation. To do that you need another way to execute jobs and handle the results. And of course there is one. Let's look at how to call stuff from the outside world.

The command line interface

While you get the most out of the development part of Databricks by using the user interface, you probably want to do a lot of the administrative parts elsewhere. The way to do that is to use either the command line interface (CLI) or the application programming interface (API.) Using them you can start clusters, list files on the Databricks File System, and a lot more.

While you can use both of them for most tasks, you'll probably end up using the CLI for quick, one-off things. If you want to automate flows, however, it's frequently easier to use the API.

We'll talk more about the API in the next chapter of the book. Right here and now, we'll focus on the CLI. It's an easy way to communicate with your workspaces from your client machine.

Setting up the CLI

The Databricks CLI works on top of Python, and you use the pip command to install it. Before you do that, however, you need an access token. We talked about this in more detail in the last chapter, so let's just go through the motions here. Note that you'll need to use your login information in the community edition of Databricks.

Click the user icon in the upper-right corner, and then click User Settings. Click the Generate New Token button and type in For CLI in the comment field. Generate the token and copy the information.

Next up, you need to install the application. At the command line or terminal, type in the following command. Note that you need to have a version of Python installed. If you don't, this won't work:

```
pip install databricks-cli
```

You'll get a bunch of dependencies installed into your system, and after a little while, you'll also have the program ready. Time to start the configuration. You do that by running the following command:

```
databricks configure
```

This will require you to respond to a couple of questions. First, you need to specify which Databricks host you are using. It'll probably look something like https:// westeurope.azuredatabricks.net/. Next is your token. That's all you need.

The result is stored in the .databrickscfg file. If you want to keep multiple of them, you can point to the relevant one with the environment variable DATABRICKS_CONFIG_FILE. Alternatively you can use DATABRICKS_HOST and DATABRICKS_TOKEN to override the configuration file.

To make sure the installation and configuration work, let's try to just list the clusters available in the workspace. Note that you can run these commands without any cluster running:

```
databricks clusters list
```

If everything works, you should get a list of all the clusters in the workspace. The first column shows the cluster ID, the second is the name, and finally you get the state. Of course this is not just for getting information; you can also trigger actions:

```
databricks clusters start --cluster-id 1111-123456-datab123
```

The cluster ID is the value you got in the preceding list. This command will tell Databricks to start the cluster for you. If you run the list command again, you'll see that the status changed. It'll go to Pending before it starts. If you don't want it to start, kill it with delete:

```
databricks clusters delete --cluster-id 1111-123456-datab123
```

This command will not delete the configuration but rather just remove the actual cluster. If you run the list command again, you'll see that the cluster is still there and that the state is Terminated.

Running CLI commands

There are a lot of different commands you can use in the CLI palette. You can get information about what's available by just running the command databricks. This actually works in lower levels as well. If you want to see what cluster commands are available, run

```
databricks clusters
```

Don't be afraid to just type in the command. If there is some argument missing, Databricks will tell you what is expected. You can also add an –h to the end of the command to get some additional information about it.

There are a group of core features that you can play around with. Clusters is what we've looked at. It gives you an opportunity to interact with the engines. Next up is fs or the Databricks File System.

Groups is for managing the security groups. We'll get to them later in this chapter. Jobs lets you add, list, and remove jobs. Libraries can be used to install and uninstall libraries onto the workspace.

Runs is what the jobs result in. So you get access to statuses from all job executions. Secrets keeps track of your keys and is also something we'll get back to in this chapter. Workspace gives you access to objects in the Databricks workspace.

Then we also have Stacks. At the time of this writing, this feature is in beta mode, and it'll probably change. I leave it out here, but it might become something good down the line, so it's worth keeping track of.

Creating and running jobs

Now, let's try a number of commands to see how they work and what you can do with them. We can start by creating a job. The easiest way to do that is to actually borrow the metadata from an existing job. This assumes that you actually have a job created. If you haven't, make one in the user interface first:

```
databricks jobs list
databricks jobs get -job-id <a job id from the result above>
```

This will return a json output with a lot of information about the job. Strip away everything except for the settings part so you have something similar to the following information. Save it to a file called Main.json. Note that you can replace the new_cluster section with something like "existing_cluster_id": "0000-000000-apress123" if you want to use an existing cluster. Note that we don't specify any node types. Databricks will use the default, which is Standard_DS3_v2 for both driver and workers, of which there will be eight:

```
{
    "name": "Apress Run 1",
    "new_cluster": {
      "spark_version": "5.2.x-scala2.11",
      "spark_conf": {
        "spark.databricks.delta.preview.enabled": "true"
      },
```

```
    "email_notifications": {},
    "timeout_seconds": 0,
    "notebook_task": {
      "notebook_path": "/Users/robert.ilijason/Main",
      "revision_timestamp": 0
    },
    "max_concurrent_runs": 1
  }
}
```

We basically stripped away all of the data that is automatically created when you add a new job. The information in settings is what you can change. So feel free to change the name, the notebook you want to run, or any other setting. Just make sure your alterations are ok. Don't reference a notebook that doesn't exist, or the next command will fail:

```
databricks jobs create -json-file Main.json
```

Run this command and you'll get a job ID back. If you go back into the Databricks user interface, you'll find your new job there. This way of creating jobs is much easier than to manually create them. Now let's run the job we just created:

```
databricks jobs run-now --job-id <your job's ID>
```

This will return the run ID. When you instantiate a job, the actual work done is called a run. The job is just the template. So to track what's happening, you need to use a different set of commands. Let's try it:

```
databricks runs get --run-id <your run's ID>
```

This command will return a lot of metadata around your job run. Most importantly you can pick up the current state. This is where you'll see if it finished ok or if it failed. You'll also be able to track it while it's running as this information is being updated live.

Accessing the Databricks File System

Next up, let's look at the file system. We talked early on about moving data back and forth from your client to the Databricks File System. Now we'll introduce yet another way of doing that – plus a little bit more. Let's start by just looking at the contents:

```
databricks fs ls
databricks fs ls dbfs:/FileStore
```

This is rather nice, right? You can look at all the files in the file system almost like in an ordinary file system on a Linux system. It's also possible to just reverse the order of the arguments and copy data from DBFS to your client. Next up, let's copy some data from the client onto the Databricks File System:

```
databricks fs mkdirs dbfs:/tmp/apress
databricks fs cp rawdata.xls dbfs:/tmp/apress/somedata.xls
databricks fs ls dbfs:/tmp/apress/somedata.xls
databricks fs mv dbfs:/tmp/apress/somedata.xls dbfs:/tmp/apress/rawdata.xls
databricks fs rm -r dbfs:/tmp/apress
```

First, we create a new subfolder underneath the tmp folder. We then copy a file from your computer into the new subfolder. We then rename it using the move command and finally remove the subfolder. The –r argument removes everything in the folder recursively.

Now there is one more interesting detail about using this fs command. By looking at the right place, you can actually see what tables are available in the Hive metadata store. Let's see how that works:

```
databricks fs ls dbfs:/user/hive/warehouse
```

This will give you a list of all tables in the default database plus all the other databases, suffixed with db. If you want to see the tables in a particular database, just add the name to the end of the ls command. While you can't look at the data this way, it's actually a nice way to scan what's available without having to spin up a cluster.

Picking up notebooks

One thing that is very relevant for development is notebook handling. Even if you do all your coding in Databricks, you probably want to move code around. Doing that manually isn't the best of ways.

While you can export and import your notebooks in the user interface, it's much easier to do in the CLI. Let's do a test to see how easy it is. We'll start by listing what's in there so we know what to export:

```
databricks workspace ls /Users/robert.ilijason/
```

Running this command, but with your name of course, will return all of your notebooks and files. While you can export the whole folder using export_dir, we'll just do one single file in this test:

```
databricks workspace export /Users/robert.ilijason/Main .
```

This will copy the notebook Main from my folder in Databricks to the current folder. It's the single dot that defines the current directory. You can use any path you want there instead. The file will be named Main.py. Feel free to look at the file in a notebook.

With this file you could easily check into a git repository or any other versioning tool. For now, let's just re-import it under a different name. It's almost as easy as exporting it, although you need to add an argument:

```
databricks workspace import --language PYTHON ./Main.py /Users/robert.ilijason/M2
databricks workspace ls /Users/robert.ilijason/
```

The language parameter is required. The accepted values are SCALA, PYTHON, SQL, and R. Using these commands, you can easily copy code between different environments and even versioning control systems if you wish.

Keeping secrets

Even if you don't plan on using the CLI for anything else, you'll pretty much be forced to do it for one feature. Secrets are the built-in way for you to use secret passwords, tokens, and other literals in your notebooks without anyone being able to see what they are.

The idea is that you store your clear text password in a vault with a tag. When you need to use the password, you ask the vault to insert it for you by referencing the tag. So while the tag can clearly be seen, no one will know the actual password.

All secrets have to be stored into a group, or a scope. You can have up to 100 groups in each workspace and a lot of secrets in each. So you will most likely have enough to get around. Let's create a scope and a secret to see how it works:

```
databricks secrets create-scope --scope apress
databricks secrets write --scope apress --key mykey --string-value setecastronomy
```

The first command creates a scope called apress. This is the group where you'll store your keys. Next, we create the actual key. We tell Databricks which scope it should store it in, the name or tag of the key, and finally the actual value. If you have a binary file you want to store, you can reference it using –binary-file /path/to/file instead of –string-value.

Before we try using the key, let's make sure it's actually there. Unless you got an error message, it will be, but knowing the commands is helpful. So let's look at the available scopes and keys within our own scope. The following commands are pretty self-explanatory:

```
databricks secrets list-scopes
databricks secrets list --scope apress
```

Ok, now that we know they are there, we can try them out. As we can't really use them for anything real, we'll have to just look at the commands. Getting the results is not possible through the CLI though, so you have to open up a Python notebook:

```
dbutils.secrets.get('apress','mykey')
```

Ah! Databricks just blocked you from seeing the secret password. Information redacted. That's the thing. You'll never be able to get it back out through the system. It's a one-way street. The way you use it is as an argument to another call. Here's an example connecting to the Azure Blog Storage using a SAS key:

```
url = "wasbs://mycontainer@mystorageaccount.blob.core.windows.net/"
configs = "fs.azure.sas.mycontainer.mystorageaccount.blob.core.windows.net"
dbutils.fs.mount(
 source = url,
mount_point = '/mnt/apress',
 extra_configs = {configs: dbutils.secrets.get('apress','mysaskey')})
```

The first two lines are information needed for connecting to Azure Blob Storage. Next, I run a command to connect to the disks. The key part is that I use a shared access signature to attach to them. That's a key that I don't want in the wrong hands. So I've created a secret and use it here. The command will work and no one can copy the key.

While you will probably work with keys mostly from the CLI, it's possible to list what's available using the built-in package. Let's list the scope and the available secrets just as we did in the CLI, but this time from Databricks:

```
dbutils.secrets.listScopes()
dbutils.secrets.list('apress')
```

Having secret keys just lying around is never a very good idea. Once you don't need them anymore, you should get rid of them. This is also done through the CLI. Let's clean up after ourselves by removing what we just created:

```
databricks secrets delete --scope apress --key mykey
databricks secrets delete-scope --scope apress
```

Small note on secrets: The cloud providers and external parties have alternatives to this solution that can be used more broadly. If you need this type of feature in many different tools, you might want to look into using them instead. If you only need this in Databricks, the built-in feature works well.

Secrets with privileges

If you want control who can call the secrets and not, you can use Access Control Lists (ACLs). This requires you to actually have access to the security features that are available as an option in Databricks.

The three levels you can set are MANAGE, WRITE, and READ. If you don't want the user to have any access at all, you just delete them from the ACL. That means they'll have no way to get to your secrets.

Let's create a small example so you can see what this looks like. Note that I assume there's a test.user user in the system. You can either create it or change the text of your own accounts. It doesn't really matter:

```
databricks secrets create-scope --scope acl_scope
databricks secrets put-acl --scope acl_scope --principal test.user
--permission MANAGE
databricks secrets list-acls --scope acl_scope
```

This creates a scope, just like before. We then add a rule that gives test.user full access to the scope. The last command will just list the ACLs so that you can see if the second command actually worked. Let's remove the user and look at the difference:

```
databricks secrets delete-acl --scope acl_scope -principal test.user

databricks secrets list-acls --scope acl_scope
```

We didn't create a secret here, but you do it just as you would in a non-ACL scope. So just use the put command and you'll be set. The same goes for other commands. It doesn't really matter if you use ACLs or not.

Revisiting cost

We already talked about the cost of Databricks and all the different models that are available on AWS and Azure. Now it's time to revisit the topic as it's pretty relevant in regard to running jobs.

As you might recall, there are a few different levels of workload types and SKUs. You have Data Engineering Light, Data Engineering, and Data Analytics. By running notebooks and other code in different ways, your jobs will be flagged differently.

"So what's the difference?" you might ask. Well, at the time of writing, the cheapest option (Data Engineering Light with standard features) is almost eight times cheaper than the most expensive option (fully featured Data Analytics). That can quickly add up to a ton of money. Here's a hot tip – don't run jobs interactively if you don't have to.

The way you get to the cheaper options is to run your code as jobs on automated clusters. That's it. Once you've developed your code, create a job and run that one instead. Don't use one of your interactive, pre-created clusters.

If you really want to run stuff cheaply, you should only run jobs with libraries. This takes you to Data Engineering Light. No notebooks. This will bring you to the cheapest load type. If you then also run it on a standard workspace with no security features, you'll be running pure Spark in a managed and cheap way.

That said, if you only want the very basic stuff, you might not need Databricks at all. So in the real world, most things you run will probably be on Data Engineering or Data Analytics. But even between those, the difference in price is huge.

Another thing is to consider the standard option. It's easy to just go with the premium version as it gives you a lot of goodies. Some might even feel it's absolutely necessary. Think twice though. Maybe you can figure out a way around the security issues by using a smart workspace strategy. At least you might be able to lower the costs for some parts.

Users, groups, and security options

With the top tier of Databricks in place, you have the option to handle users, groups, and security on a very granular level. This will probably be what you use mostly in an enterprise setting, unless you have clever administrators and insensitive data.

If you don't have the higher tier of functionality, you won't be able to do any kind of detailed management. All users will have full access to everything. It is a bit extreme, and clearly Databricks wants you to buy all the expensive options.

Users and groups

The core setup in Databricks is based on users and groups. If you're an administrator, you'll have access to both. Click the half-human icon in the upper-right corner and click the Admin Console link.

Under the first tab, you'll see a list of users that have access to the environment. It's probably just you right now. To the right of the name, you'll see if the user has administrative privilege and is allowed to create clusters. In a non-high-end version of Databricks, both will be checked for all users. At the far end, there's also an x. That's for removing the user.

While it's possible to manage every user individually, it's usually a good idea to group users together. You might, for instance, want to have a group for administrators and another one for developers or data scientists. That makes things much easier to handle.

Let's start by creating a couple of users. To do so, click the Add User button, and you'll get a box where you type in the email address to the new user. There might be limitations to what addresses you can use, depending on what version you use. You can override them if you want to do so for testing purposes although the users might not be able to log in.

Now we should create a group and add our users to it. Go to the Groups page and click the Create Group button. Name the group Apress and click Create. This will bring you to a screen where you add users and map them to privileges.

Add our new users to the members list using the Add users or groups button. Note that you can nest groups within groups. When you've done this, click the Entitlements link to see what privileges you can give them already here. Feel free to enable Allow cluster creation if you want to.

Now you have both users and groups ready. Time to start looking at how to use them. To do that, you'll have to enable features that are available under the Access Control tab. So let's see what's available there.

Using SCIM provisioning

One small note: If you already have a provisioning solution set up in the cloud, you might want to use that directly instead of handling privileges. Databricks can use the System for Cross-domain Identity Management, SCIM, solution for Azure Active Directory, Okta, and OneLogin.

The core idea is that you handle the access elsewhere. I highly recommend this as it lowers the risk of users just lingering after they left the company. Also, if you have external resources, you can handle all privileges in one place.

Setting this up isn't very hard, but outside the scope of this book as most of the work is done in the external tools. You can search for SCIM in the official Databricks documentation to find step-by-step instructions.

Access Control

Under the Access Control tab, you have four options. If you pay for the full version, they'll all be enabled by default. Unless you have good reasons to do otherwise, you should keep it that way. If you don't have the top-end version, you have fewer options.

As mentioned before, if you don't feel you have the need for any of the features offered, consider switching to a cheaper type of workspace. There is no reason to pay (a lot) more for things you don't plan on using.

That said, what's offered is pretty good and in many cases needed. There are security issues you have to take into consideration with the basic workspace. I'll keep the explanations short as they are mostly self-explanatory.

Workspace Access Control

Databricks is a collaborative environment that makes it easy to share code between developers. That said, a lot of people like to have their own, personal files that they do not share with others. To enable private files, you need to enable Workspace Access Control. It is activated by default.

Everything in the personal folders in the workspace stays private. Files stored in Shared are available to everyone. Users can grant others access to their files. The default is no access, but you can set it to Read, Run, Edit, and Manage. To do so, find the file in your folder, right-click, and select Permissions. Just add and remove access as you see fit.

I personally think sharing should be done as much as possible. It enables collaboration in a much better way. Developers have a tendency to keep their code hidden until they think it's perfect. That might be very late in the project, and mistakes that were easily fixed two months ago can be costly to fix now. Be as open as possible.

Cluster, Pool, and Jobs Access Control

This feature helps you limit access to clusters and pools and access to job information. It basically lets you lock things down and give users and groups permission to use the different resources.

If you want to define who has access to a cluster, open the detail page for it and click the Advanced Options. The last tab in the new row will be Permissions. Here you can tell who has the privilege to Attach, Restart, and Manage it.

For jobs, go to the detail page and expand the Advanced option. There is a Permissions option at the bottom of the new list. Click the Edit link, and you'll get the same dialog box as you've gotten used to now. Just select who'll be able to view it and manage runs.

A small note: If you let anyone create and run clusters freely, you need to have good processes in place. Setting up a huge cluster is easy. Paying the bill at the end of the month is not. So if you let people manage this themselves, make sure they know their boundaries.

Table Access Control

This is probably the feature that you'll have to use most if you want to shield certain parts of the system from some users. It enables you to give access on a table and, indirectly, even specific columns to users and groups.

There are, however, a few drawbacks. First of all, your cluster has to be of the High Concurrency type. You also have to check a box under the advanced options to enable this feature. If you do that, you won't be able to run either Scala or R on the cluster. Oh, at the time of this writing, the Python support is in beta. It's been in beta for good long time. Hopefully it isn't any more when you read this.

So if you want proper, solid, well-tested support for this feature, you'll have to lock it down to SQL only. To do that, add this line in the Spark Config box under the Spark tab in the Advanced Options section:

```
spark.databricks.acl.sqlOnly true
```

Whether you go with just SQL or add Python to the mix, you'll get the ability to grant users and groups privilege on tables. You can grant SELECT, CREATE, MODIFY, READ_METADATA, CREATED_NAMED_FUNCTION, and ALL_PRIVILEGES using the GRANT command. This is what it looks like:

```
GRANT SELECT ON DATABASE apress TO 'robert.ilijason';
GRANT SELECT ON apress.mytab TO 'robert.ilijason';
GRANT ALL PRIVILEGES ON apress.myview TO 'robert.ilijason';
REVOKE SELECT ON apress.myview FROM 'robert.ilijason';
REVOKE ALL PRIVILEGES ON apress.myview FROM 'robert.ilijason';
```

These commands are self-explanatory. You either GRANT or REVOKE privileges to and from a user. You can do it on a high level, like a complete database, or on a lower level, like a table. While all examples are with a user, groups also work in the same way.

If you want to restrict a few columns in a table, you can create a view on top of the table without the columns you don't want to expose. We can do a small experiment to see how this works:

```
CREATE TABLE X (notsecret integer, secret integer);
CREATE VIEW X_view AS SELECT notsecret FROM X;
GRANT SELECT ON X_view TO 'untrusted.employee';
```

Then as the untrusted user, you can try to select data from the table and view. You will have access to the view, but not the underlying table. While this is a good feature to have, I think it's a bit clunky to use. That's especially true for tables you change the structure of frequently:

```
SELECT * FROM X;
SELECT * FROM X_view;
```

There's another trick worth mentioning that might make your life easier. With the DENY command, you can explicitly forbid a user from accessing, for instance, a table. That's helpful if you want to give someone access to all tables in a database except for one:

```
DENY SELECT ON X to 'semi-trusted.employee';
```

Compared to modern relational database management systems, Databricks is weak in the table access part. That said, you have a lot of options to handle this in a different way. You can use different workspaces and external solutions. On Azure you can, for instance, enable credential pass-through to the data lake storage solution.

Personal Access Tokens

Earlier in this chapter, we played around with the CLI. In other chapters, we've connected to Databricks using external tools. In both cases we've used a token for authentication. This is a nice and simple way to access the environment from the outside.

It can however be a security weakness. The only thing you need is one key, and you can access most of the system. There might be situations where you don't want that to be possible. That's when you set the Personal Access Token to Disabled.

The default setting is Enabled, and while it might make sense to shut it down in some cases, it also makes working with the system harder. It's not only the CLI that stops working but also the API and all other external calls using tokens.

The rest

There are quite a few other features in the tool as well. You can purge storage, remove cluster logs, and enable container services. The stuff in workspace storage doesn't require much explanation.

The Advanced tab gives you a handful of options that you should leave as is unless you have very good reasons to do otherwise. They are there to help you secure your environment. Think twice before changing them.

There is one thing you should be aware of in the advanced section. Enabling X-Frame-Options Header blocks third-party domains from putting Databricks in an Iframe. This is enabled by default and should normally stay that way.

It might be that you want to put Databricks into an Iframe on your company Web. If so, you might need to flip this switch. I don't think it's a good idea to do this, but if it ever comes up. you know where to look.

Summary

In this chapter, we looked closer to features that you need to administer the system. Before that, we listed a few good things to keep in mind when thinking about putting code into production.

We then spent quite a bit of time in the jobs section, looking at how we can start notebooks automatically in the background. We learned about how to pass arguments and also how to create a scheduled run.

Next, we dove into the wonderful world of the command line interface. We looked at how we can execute a lot of the commands in Databricks from a terminal prompt on our client. This enables us to use external tools for things like running jobs.

Finally, we looked at users, groups, and the security features that Databricks offers to the ones that pay up. We learned how to restrict access to things like clusters, tables, and even creation of tokens.

CHAPTER 11

Bits and Pieces

We've covered a lot of the basics in the book so far. But there are a lot of things we haven't touched upon or dove into enough. In this chapter, we'll look at a few things that are worth looking at so you know what's available.

To start with, we'll look at machine learning. It's the topic de jour, and you might imagine that Apache Spark is very well suited for it. We'll take a quick look at MLlib, a library that is especially well suited for running these types of load on Databricks.

Next, we'll look at a Databricks-backed feature called MLflow. It's a functionality that will make it easier for you to keep track of your machine learning experiments. While it is open source, it's very well integrated with Databricks.

After that it's time to show an example of how you can handle updates with Parquet files without the help of Delta Lake. This is not as trivial as one might wish. Once you know the core flow, it'll be breeze though.

Then we'll introduce another library that Databricks supports. It's called Koalas and can be the bridge to Apache Spark that many single-thread Python developers want. It lets you run commands like if you used the popular library pandas.

After a lot of data, it's then time to look at what we can do with the presentation layer. While there are other tools that do this much better, you have some basic functionality to deliver dashboards for your user.

We also look at how you can play around with a lot of the features in Databricks with code, using the API. You'll learn how to start clusters, run jobs, and a lot more without having to log in to the graphical user interface.

Finally, we'll look at another way of handling data flows. Instead of running bulk jobs, we'll look at how to use streaming. We'll see how we can ask Databricks to process information as it comes in.

That's a lot, so let's get going.

CHAPTER 11   BITS AND PIECES

MLlib

Machine learning is a common use case on Apache Spark. Running algorithmic analytics on a large dataset is taxing on the processing system, so it makes sense to spread it out across many machines.

We've mentioned earlier in the book that there is a library with machine learning functions especially adapted for Apache Spark. It's called Machine Learning library, or MLlib for short, and is maintained by the same community that's supporting the main product.

While it might not be able to compete with R on the width of algorithms or with scikit-learn on single-machine performance, it works well when using common algorithms on large datasets.

There are some options for most common use cases in the library. You can run regressions, classifications, recommendations, clustering, and more using MLlib. Let's use a simple one as an example: Frequent Pattern Growth.

Frequent Pattern Growth

One common analysis in both the ecommerce and retail world is helping customers find things they might like, based on what they've bought in the past. To do that you need to analyze all the receipts, or transactions.

There are a few ways to do this, but one common technique is called a market basket analysis. It uses association rules to figure out how frequently items occur together in transactions. By looking at the result, you can find interesting connections that are not always obvious beforehand.

There are a number of different algorithms for doing this, like Eclat and Apriori. In MLlib you have FP-Growth. This is an implementation that is fast and well adapted to Apache Spark. It's also easy to use, as you'll see.

To make it work, you need to prepare the data in a certain way. You need a transaction id in one column and a list of items in the other. This is a bit cumbersome if you have the data in normal database tables, but we'll look at how to do this in the example.

Two input arguments are also needed for the algorithm to run, support and confidence. This is a way for you to limit the output. Without it you'd get a lot of results back if your dataset is large. So let's go through what they mean.

The support is calculated by counting the number of transactions where an item set is represented, divided by the total number of transactions. This can be one item or several. So if item A and item B appear in five transactions out of ten, the support is 5/10 = 0.5:

AB, **ABC**, **ABD**, BD, BCD, **ABD**, AD, ACD, **AB**, AC

Confidence tells you how strong the connection between two products is. It compares how frequently we see this association in the dataset to how frequently item A shows up:

AB, **ABC**, **ABD**, BD, BCD, **ABD**, AD, ACD, **AB**, AC

In our example we've already seen that the AB combination is available five times. Item A is available on its own eight times. That gives us a confidence of 5/8 = 0,625. The higher the value, the more likely it is that item A is together with item B.

Confidence is important. Consider, for instance, a shopping bag. It'll give you a high support for almost any product as most people will buy it. The confidence will however be low as it's occurring in almost every transaction.

Creating some data

To run our tests, we need some data. Let's keep it simple so you can easily track what's happening: one table for the products and one for the carts. As I'm thirsty while writing this, we'll go with soft drinks:

```
create table products
        (product_id integer, product_name string);
insert into products values (1, 'Coca Cola');
insert into products values (2, 'Pepsi Cola');
insert into products values (3, 'RedBull');
insert into products values (4, 'Evian');

create table cart (cart_id integer, product_id integer);
insert into cart values (1, 1);
insert into cart values (1, 2);
insert into cart values (1, 3);
insert into cart values (2, 1);
insert into cart values (3, 2);
insert into cart values (3, 3);
insert into cart values (3, 4);
```

```
insert into cart values (4, 2);
insert into cart values (4, 3);
insert into cart values (5, 1);
insert into cart values (5, 3);
insert into cart values (6, 1);
insert into cart values (6, 3);
insert into cart values (7, 2);
insert into cart values (8, 4);
insert into cart values (9, 1);
insert into cart values (9, 3);
insert into cart values (10, 1);
insert into cart values (10, 3);
insert into cart values (10, 4);
```

The first table, products, contains four soft drinks that our imaginary store is selling. Cart lists all the transactions. We'll combine them in the next step to make the data ready for FP-Growth.

If you want to do this in Python and avoid the table creation altogether, you can create the DataFrame directly. Here's an example of how you can do that with the parallelize command, but only for the first two rows:

```
from pyspark.sql import Row
rdd = sc.parallelize([
Row(cart_id=1,products=['RedBull','Coca Cola','Pepsi Cola']), Row(cart_id=2,products=['Coca Cola'])])
df = rdd.toDF()
display(df)
```

Preparing the data

As mentioned earlier, the FP-Growth algorithm requires data in a given format: one column for a transaction id and another one with all the items. Let's build a DataFrame that returns exactly what it needed:

```
carts = spark.sql('select p.product_name, c.cart_id from products p join cart c on (c.product_id = p.product_id)')
display(carts)
```

```
from pyspark.sql.functions import collect_set
preppedcarts = carts
        .groupBy('cart_id')
        .agg(collect_set('product_name').alias('products'))
display(preppedcarts)
```

The first part is nothing new. It's just a join between the cart and the product table. This will give us a long list of our items and in which carts they appear. This is not what the FP-Growth algorithm wants though.

In the next step, we use the collect_set function to pull all the data from product_name and turn it into a list. The lists are grouped by the cart_id. This will give us the format we need. Just for a reference, there's a collect_list function as well, but collect_set removes duplicates.

Running the algorithm

Time to run the algorithm. As you'll see, there isn't much to it, just a couple of lines to do the actual work. This is actually true for most machine learning work. The code to run the actual machine learning parts is usually much smaller than the preceding cleanup code:

```
from pyspark.ml.fpm import FPGrowth

fpGrowth = FPGrowth(itemsCol="products"
        ,minSupport=0.5
        ,minConfidence=0.5)

model = fpGrowth.fit(preppedcarts)
```

We start by importing the FPGrowth library. Then we call the function. We tell it what column contains the products and set both the support and confidence levels. With that we have the settings done, but we also need to run it.

Executing it we do by using fit. This command will go through the actual data and start counting. In our case it's an operation that takes seconds. If you have a few billion rows, it can take a while, especially if you have aggressive support and confidence settings.

CHAPTER 11   BITS AND PIECES

Parsing the results

Once it's done it's time to look at the results. We want to know what information we can get out of this dataset. There are three main things to look at: a frequency table, the association rules, and some predictions.

The first box shows you the frequency of items and combinations of items. As we have a total of ten transactions and a support of 0.5, we'll get everything that occurs in half of the time:

`display(model.freqItemsets)`

The only pair we get is Coca Cola and Red Bull that shows up in five transactions. So it seems that persons who buy one are also interested in the other. Then on the other hand, both are popular overall.

Next up is a box showing us the confidence and lift for the multi-item lines. It shows the results from both perspective. Antecedent is the base item and consequent the following item in the analysis:

`display(model.associationRules)`

As you can see it's more likely that you buy a Red Bull if you bought a Coca Cola than the other way around. The actual difference in our case is very small though as our dataset is as well.

Then there's lift. Lift tells us how likely it is item B is bought when item A is, taking into consideration how frequently item A is bought. A value above one means it's likely. A value below one means it's unlikely.

To get the value, you take the support for the item set and divide it with the support for the antecedent times the support for the consequent. It sounds more complicated than it is.

So let's see. Our supports are 0.6 (6/10) for Coca Cola and 0.7 (7/10) for Red Bull. Combined they give us 0.6 × 0.7 = 0.42. The support for the combination of Red Bull and Coca Cola is 0.5. So the lift is 0.5/0.42 = ~1.19.

Finally we get a list of all transactions with a prediction. It basically uses the associations we've found and adds the items that are missing to fulfil the expected. So you get a Coca Cola prediction for all rows with Red Bull missing the sugary soft drink and the other way around:

`display(model.transform(preppedcarts))`

Note that it can be hard to know for sure if the stuff you find is relevant. It's easy to jump to conclusions and make erroneous assumptions. Like always, try to validate what you find before you take decisions based on the results.

MLflow

When working with machine learning algorithms, keeping track of what you do is crucial. Doing it manually is hard and prone to errors. Ideally you want all the in-parameters and results to be stored for future reference, together with the code if possible.

There are a number of solutions to do this. One of them is built by Databricks and released as open source. The name is MLflow and as you might expect it's deeply integrated within the Databricks suite of tools.

MLflow is useful for both tracking and project packaging. It also offers a general format for models. In this chapter we'll focus on the first part, so you get an intuition for how to use it in your machine learning experiments.

Let's look at how you can use this feature in a normal workflow. We'll pick a prepared dataset, run a regression, and randomly set parameters. Each run will be stored into the MLflow database. In the end we use a built-in graphical user interface to quickly see all the results.

Running the code

First of all, we need to import a couple of libraries. We need, of course, MLflow. On top of that, we'll use the popular machine learning library scikit-learn. It's a popular tool used by a lot of data scientists.

It also offers a handful of prepared datasets to play around with, optimized for machine learning tests. We'll use housing data to see if we can predict price based on all features. It's not the way to go, but the important thing here is the MLflow parts:

```
dbutils.library.installPyPI('mlflow')
dbutils.library.installPyPI('scikit-learn')
dbutils.library.restartPython()
```

With the libraries imported, it's time to import the data. We'll get the data from scikit, turn it into a pandas dataframe, and then split it into two Apaches Spark DataFrames. One of them we'll use for training and the other one for testing. The latter one will contain about 20 percent of the data.

Note that you should pull random rows in a real scenario and possibly add a validation set. For small datasets like this, you would probably want to use cross-validation to minimize the risk of overfitting:

```
import pandas as pd
import sklearn
from sklearn.datasets import load_boston

basedata = load_boston()
pddf = pd.DataFrame(basedata.data
        ,columns=basedata.feature_names)
pddf['target'] = pd.Series(basedata.target)

pct20 = int(pddf.shape[0]*.2)

testdata = spark.createDataFrame(pddf[:pct20])
traindata = spark.createDataFrame(pddf[pct20:])
```

We end up with two DataFrames, testdata and traindata, which we'll use in a bit. First, we need to create a structure that the MLlib regression algorithm can use. We'll do that using the VectorAssembler function:

```
from pyspark.ml.feature import VectorAssembler

va = VectorAssembler(inputCols = basedata.feature_names
        ,outputCol = 'features')
testdata = va.transform(testdata)['features','target']
traindata = va.transform(traindata)['features','target']
```

VectorAssembler takes the columns defined in inputCols, which in our case are all columns except for target, and turn them into a list. So if you look at the DataFrame, you'll notice that it now has two columns, features and target. The first one contains a list of all dimensions we want to use. The second one is the house price.

So with all the features combined, we have a numeric value to predict – a perfect case for a regression algorithm, although we should probably have pruned a few columns in reality. Anyways, let's run the actual code to see how well we can predict the values.

To get some test data for MLflow, we'll run it three times using a simple loop. We'll also change the maxItem and regParam hyperparameters to see if they make any difference in the performance of the algorithm. To keep the loop logic to a minimum, we'll increase both in unison. You wouldn't do it like this in a real use case:

```
from pyspark.ml.regression import LinearRegression
import mlflow

for i in range(1,4):
  with mlflow.start_run():
    mi = 10 * i
    rp = 0.1 * i
    enp = 0.5

    mlflow.log_param('maxIter',mi)
    mlflow.log_param('regParam',rp)
    mlflow.log_param('elasticNetParam',enp)

    lr = LinearRegression(maxIter=mi
        ,regParam=rp
        ,elasticNetParam=enp
        ,labelCol="target")

    model = lr.fit(traindata)
    pred = model.transform(testdata)

    r = pred.stat.corr("prediction", "target")
    mlflow.log_metric("rsquared", r**2, step=i)
```

This wasn't so hard. For each round, we explicitly tell MLflow that we're starting a new run. We then set the parameters, fit the data, and use the model on the testdata to get predictions. Finally we check how well we guessed, using the R-squared value.

The problem is that we ran this multiple times, got different values, and don't really know what parameter values gave us the best results. Luckily we asked MLflow to track them for us. We told it what parameters we used with the log_param command and the results with log_metric. We could also have saved the model or an artifact to be stored with the run. There are a lot of options. Time to see what this actually looks like in the user interface.

CHAPTER 11 BITS AND PIECES

Checking the results

Let's start by clicking the Runs link in the upper-right corner of the work area. This will reveal all the parameters and metrics for your runs. You can expand each row to get the information in a row-based format, making it easier to read.

So you can look at all these values and figure out what run was the best. Not only that, a nice feature is that you can ask Databricks to show you the version of the notebook that was run for a given result. That means you can just play around with a lot of options without keeping much track and then go back and find the code that worked the best.

Now, if you click the link button in the upper-right corner, next to the reload button, you'll get to a totally new user interface. This is an overview of all your runs, with the ability to sort and search using simple clicks on the mouse.

You even have the option to run queries on the data. If you, for instance, want to see all results where the performance for a given parameter is above a certain threshold, you could type in something like this:

```
params.maxIter = 20 and metrics.rsquared > 0.65
```

As you define the parameters and metrics yourself, you can do really nice things here, quickly filtering out data and finding what works. As mentioned there's an API you can play around with to build logic around these values. For instance, you might want to retrain a model once the performance falls below a threshold too see if it improves.

Overall the MLflow view is where you can look through hundreds of models and see how they perform. You can investigate how different settings affect the result, as we just did, or track the results over time. A model that works well today might not be good tomorrow.

Updating tables

If you decide to store data externally in Parquet format without using Delta Lake, you'll notice that it's not trivial to update the underlying data. If you have small tables, it doesn't really matter as you can read and replace them every time. With tables in the terabytes, it might be more cumbersome to use the replace method.

There are a number of ways of doing this, including just hard-replacing the underlying Parquet files. If you want to do it somewhat structured though, you can use a little trick. It's not straightforward, but not too complicated either.

It should be mentioned once again that Delta Lake is probably the way to go here if it works in your environment. There are both pros and cons with this Databricks-developed, open source solution. In general it's something well worth considering.

Create the original table

To begin with we need a source table. We'll create one with underlying partitions as well to make it clearer, but it's not necessary. For the same reason, we'll keep the data to one row per partition.

In this example I'll assume you have access to an Oracle database. The solution is similar if you use SQL Server, DB2, SAE, or any other relational database, but you might need to do some changes in the code.

Log in to the Oracle database and run the following SQL. It'll create the source table. Make sure to keep track of which schema you create it in. You'll need to reference it when you read it from Databricks later:

```
drop table p_table;
create table p_table (
a date, b varchar2(10))
partition by range(a)
(partition p1 values less than (to_date('20200102','YYYYMMDD')),
partition p2 values less than (to_date('20200103','YYYYMMDD')),
partition p3 values less than (to_date('20200104','YYYYMMDD')),
partition p4 values less than (to_date('20200105','YYYYMMDD')),
partition pm values less than (MAXVALUE));

insert into p_table values (to_date('20200101','YYYYMMDD'),'Data 1');
insert into p_table values (to_date('20200102','YYYYMMDD'),'Data 2');
insert into p_table values (to_date('20200103','YYYYMMDD'),'Data 3');
insert into p_table values (to_date('20200104','YYYYMMDD'),'Data 4');
commit;
```

This will give us four rows, one per date. Every row has a corresponding text. Note that we're using the Oracle date format. JDBC will try to convert this into timestamp if we don't tell it not to. Let's move to the next step where you can see how we do that.

CHAPTER 11   BITS AND PIECES

Connect from Databricks

Next up, we'll connect to the Oracle database from Databricks to read the table for the first time. Note that you need to have port openings and the Oracle drivers in place. I won't go into this in detail here as we've discussed it earlier in the book.

In the following command, we use Oracle JDBC to connect. As you see we set the lowerbound and upperbound to make partitioning work. We set the partitions to four as we know that's the number we have and tell Databricks which column to slice by on reading. Don't forget to add the schema name to the SQL if you log in with another user:

```
df = spark \
.read \
.format("jdbc").option("url", "jdbc:oracle:thin:@//<servername>:1521/<servicename>") \
.option("dbTable", "(SELECT * FROM p_table)") \
.option("user", "<username/schema>") \
.option("password", "<password>") \
.option("driver", "oracle.jdbc.driver.OracleDriver") \
.option("fetchsize", 2000) \
.option("lowerBound", "2020-01-01") \
.option("upperBound", "2020-01-04") \
.option("numPartitions", 4) \
.option("partitionColumn", "a") \
.option("oracle.jdbc.mapDateToTimestamp", "false") \
.option("sessionInitStatement", "ALTER SESSION SET NLS_DATE_FORMAT = 'YYYY-MM-DD'") \
.load()
```

The only interesting part here are the two last options. By setting the mapDateToTimestamp to false, we explicitly tell JDBC to leave the dates alone. In the next line, we send a command to Oracle, telling it to use the YYYY-MM-DD date format. Ok, time to get the data:

```
df.write.partitionBy('A').parquet('/tmp/ptab')
```

Now we have the core table cloned from the database to our local storage. In this case we're using DBFS, but in reality you'd probably use some external option instead. Either way, it's time to create some changes and pull some data.

To look at how the data is stored, you can like always use the built-in commands to look at the file system. You'll notice that the partition key is stored as a folder name, which will be relevant later in this operation:

```
%fs ls /tmp/ptab
```

Pulling the delta

To get some delta we need to create it first. We do that by updating a row in the table. This is done with an ordinary update statement in the database. Just run the following commands in the database:

```
update p_table set b = 'Data 22'
       where a = to_date('20200102','YYYYMMDD');
commit
```

The actual delta handling can be done in many different ways. If possible, the easiest is to do it at the source. Just keep track of DML when loading data. Often this is not possible and you need to use some other technique.

In Oracle there are a number of tricks to do this, but this is outside the scope of this book. Worst case is you'll have to compare tables, which is expensive and slow. In our case we luckily know what we've done and can just pick up the correct partition:

```
df_delta = spark \
.read \
.format("jdbc") \
.format("jdbc").option("url", "jdbc:oracle:thin:@//<servername>:1521/<servicename>") \
.option("dbTable", "(SELECT B,A FROM p_table partition (p2))") \
.option("user", "<username/schema>") \
.option("password", "<password>") \
.option("driver", "oracle.jdbc.driver.OracleDriver") \
.option("fetchsize", 2000) \
.option("lowerBound", "2020-01-02") \
.option("upperBound", "2020-01-02") \
.option("numPartitions", 1) \
.option("partitionColumn", "a") \
```

CHAPTER 11 BITS AND PIECES

```
.option("oracle.jdbc.mapDateToTimestamp", "false") \
.option("sessionInitStatement", "ALTER SESSION SET NLS_DATE_FORMAT =
'YYYY-MM-DD'") \
.load()
display(df_delta)
```

This will give you a DataFrame with a diff. There are a few things we've changed here compared to reading the whole table. In the SQL we specify the exact partition we want to pick up, to minimize the data we need to read. This could have been done with a WHERE clause, but I find it easier to specify the partition if I know it.

Another important thing is that we now specify the order of the columns, with A at the end. The reason for this is that Apache Spark puts the partition key column at the end and the source and target tables have to match. Else you will corrupt your data. Feel free to try it out.

Verifying the formats

As I mentioned earlier, the source and target have to match. To avoid corruption it's wise to do some checks before you do the merge. If you know your data well, it might be unnecessary, but it's quick to run. First, let's pick up the original data into a new DataFrame:

```
df_original = spark.read.parquet('/tmp/ptab')
```

Next up, it's time to do the actual check. What we need to validate are the schemas. So we pick them up to begin with. Then we just compare them. If there's a failure, we try to find what it is and print it out:

```
orig_schema = df_original.schema.json()
delta_schema = df_delta.schema.json()

if df_schema == data_schema:
  print('The naming of source and target are identical')
else:
  print('There is a diff in here that will cause issues')

  import json
  json_source = json.loads(df_schema)
```

```
json_delta = json.loads(data_schema)
list_source = ['{}:{}'.format(f['name'], f['type']) for f in json_
source['fields']]
list_delta = ['{}:{}'.format(f['name'], f['type']) for f in json_
delta['fields']]

print(set(list_source) ^ set(list_delta))
```

This is somewhat simplified check, but will do the job. The code should be fairly obvious by now. The only new thing is the last command where we use the ^ as a shorthand for the symmetric_difference function. What it'll do is to show what's in list_sources or list_delta, but not in both – so the differences.

If the result here isn't an ok, you shouldn't continue. Look at the result you get, go back, and check what went wrong. Just redo everything if necessary. Once the schemas are identical, continue to the update part.

Update the table

Finally it's time to update the table. First, we need to create a temporary view. This is required for the next step. Next up, we need to change the partitionOverwriteMode setting to dynamic. Finally we run the command that does the writing:

```
df_original.createOrReplaceTempView('orig_v')
```

```
spark.conf.set
("spark.sql.sources.partitionOverwriteMode","DYNAMIC")
df_delta.write.insertInto('orig_v', overwrite=True)
```

That's it. The actual command is very easy once you have done all the work leading up to it. Now that we have the new table, let's create a new DataFrame and verify that the changes we've done are represented in the Parquet files:

```
df_verify = spark.read.parquet('/tmp/ptab')
```

```
display(df_verify)
```

```
%fs ls /tmp/ptab
```

A short note about Pandas

If you start browsing the net, looking for data science information, you'll notice that there are a few tools that pop up more frequently than others. Within the Python sphere, almost everyone uses the three core libraries numpy, scipy, and pandas to some extent.

It's the last one, pandas, that you'll pretty much have to give up when you move to Spark. Pandas is an easy-to-use, open source data analysis library that you can run on your client computer. Also, it's built around the concept of dataframes.

While pandas is great at what it does, scaling is not really what it's built for. You can install and run it on an Apache Spark cluster, but the work won't be automatically distributed. Instead you'll be bogged down by the memory and processing limitations of a single machine.

For this reason many developers run pandas when they're developing on their local machines, but use other solutions when the data amounts increase dramatically. This is frequently the library Dask or Apache Spark. Unfortunately the conversion from small scale to large scale isn't always trivial as the syntax isn't one-to-one.

Koalas, Pandas for Spark

For this reason a small team started looking at ways to simplify the transition between pandas and DataFrames. The way they decided to do it was to create a library which should work like pandas, but use Apache Spark DataFrames underneath the surface.

The result is called Koalas and was introduced in 2019. You can use your normal pandas syntax, but when executing code, the work will be done in Apache Spark. Ideally you can just take the code you developed with test data on you client and throw it on a cluster for crunching large datasets.

At the time of this writing, the small team has done quite a lot of work, and many things really work out of the box. Still, there's quite a long road ahead of them before this will fully work as wished.

Databricks is deeply involved in the project, for obvious reasons. A working Koalas would make the transition to Apache Spark a lot easier for companies already having a lot of code using pandas. So let's look at how to get Koalas running on your cluster and what the code looks like.

CHAPTER 11 BITS AND PIECES

Playing around with Koalas

Let's start by looking at a small example in pandas. We'll create a dictionary with some Tesla cars and convert it into a pandas dataframe. Note that this is different from an Apache Spark DataFrame. You can only use pandas functions on this one and vice versa:

```
import pandas as pd

models = {
  'Model' : ['Model S'
        ,'Model 3'
        ,'Model Y'
        ,'Model X'
        ,'Cybertruck'],
  'Entry Price' : [85000, 35000, 45000, 89500, 50000]
}
df = pd.DataFrame(models, columns = ['Model', 'Entry Price'])
df.columns = ['Model','Price']

print(df)
print(type(df))
```

Nothing fancy here. Once we've created the pandas dataframe, we change the column names and print the content as well as the type. Note that it's a pandas.core.frame.DataFrame. So it's different. Time to test this exact same thing using Koalas.

First up, we need to get the library. We do that as we do with any library. You can either add it the normal way or, like here, just install it on the fly. If you start using it more frequently, it's smarter of course to use the library functionality in Databricks so you have it available every time you start the cluster:

```
dbutils.library.installPyPI("koalas")
dbutils.library.restartPython()
```

Next up, we need to create the actual code. Like before we create a dictionary, make it into a dataframe, and change the column names. Then we just print out the data and information about the data type:

243

CHAPTER 11   BITS AND PIECES

```
import databricks.koalas as ks

models = {
  'Model' : ['Model S'
        ,'Model 3'
        ,'Model Y'
        ,'Model X'
        ,'Cybertruck'],
  'Entry Price' : [85000, 35000, 45000, 89500, 50000]
        }
df = ks.DataFrame(models, columns = ['Model', 'Entry Price'])
df.columns = ['Model','Price']

print(df)
print(type(df))
```

What you'll notice is that the code is almost identical to the one we wrote earlier. The only difference is that we imported Koalas instead of pandas and used a different alias. If you'd use pd instead of ks as the alias, it'd be one-to-one.

The result is different though – not the actual output, which is the same, but the data type. It's a databricks.koalas.frame.DataFrame. This is different from pandas, but also different from a normal Apache Spark DataFrame. So it's yet another type of dataframe. Of course you can't freely just move back and forth between command sets.

Let's look at another example. This time we'll read a couple of files, append them, and do a bit of basic aggregation. We'll see a couple of things in this example and discover there are issues:

```
df0 = pd.read_csv('/dbfs/databricks-datasets/airlines/part-00000')
df1 = pd.read_csv('/dbfs/databricks-datasets/airlines/part-00001',header=0)
df1 = pd.DataFrame(data=df1.values, columns=df0.columns)

df = pd.concat([df0,df1], ignore_index=True)
df[['Year','Month']].groupby('Year').sum()
```

You'll notice that pandas is not DBFS aware. So you need to write paths as they are seen on the driver node. Otherwise, this is pretty straightforward. Note though the way we create a dataframe on line 3. This will be important when we rewrite this for Koalas. Let's try it out:

CHAPTER 11 BITS AND PIECES

```
df0 = ks.read_csv('/databricks-datasets/airlines/part-00000')
df1 = ks.read_csv('/databricks-datasets/airlines/part-00001',header=0)
df1 = ks.DataFrame(data=df1.values, columns=df0.columns)

df = ks.concat([df0,df1], ignore_index=True)
```

Hey, that didn't work out the way we wished. Instead of the expected result, we get an error. Apparently the DataFrame.values() function isn't implemented in Koalas yet. This is sad but something you'll see quite often if you have a lot of code:

```
df0 = ks.read_csv('/databricks-datasets/airlines/part-00000')
df1 = ks.read_csv('/databricks-datasets/airlines/part-00001',header=0)

df1.columns = df0.columns
df = ks.concat([df0,df1], ignore_index=True)

df[['Year','Month']].groupby('Year').sum()
```

Here's the code slightly rewritten so that you can get it to run properly. As you can see, Koalas knows what DBFS is and the paths are different. Once we've changed the way we set the column names, the rest works just fine.

If you use Koalas a lot, you'll notice that most frequently used stuff often works as expected out of the box. It is small things like the one we just saw that don't. It can be frustrating, but still absolutely worth testing out if you have existing code.

You will notice however that the speed isn't great in our test. Quite the opposite. It isn't until you get a large amount of data that you see the benefit of using Koalas. Try running an analysis on all airlines files, and pandas will collapse. It just can't handle it, memory-wise. Koalas can.

The future of Koalas

It's always hard to evaluate how successful projects like these will be early on. There are a lot of really great projects that lie dusty by the side of the road. A few things about Koalas make me think it will survive. Other things make me think it won't. Time will tell.

What makes Koalas feasible is the huge amount of pandas code out there. If successful, this project promises companies that they can more or less just change a library import and then immediately get scale on existing code. That's a strong argument.

Another pro is that Databricks seems to back it. It makes sense for them to do it, and as the biggest Apache Spark company around, they can throw some resources at the problem. Also, they can spread the word. Corporate backing is always helpful.

The problem is that getting to full compatibility is hard. Like any projects, it's easy to pick the low-hanging fruits in the beginning, but much harder to get the last features done. Almost compatible might be good enough for some, but not most as you need the code to work as expected.

It's also crucial that this library is well supported going forward, with all changes in pandas somewhat quickly being copied into Koalas. That might happen if there's enough support for it, but it's a chicken-and-egg problem.

We'll see what the future brings for Koalas. Meanwhile, I recommend using Pyspark from development to production if you have the choice. Changing tools in the middle is never a great idea as it'll introduce risk and require more testing.

The art of presenting data

This book is focused on manipulating the data and creating results, not showing it off. While that's what the end user often sees, presenting information is a huge topic in and of itself. It's very easy to do it poorly and requires a different skill set than pulling, cleaning, and running algorithms on the data.

If you're interested in the topic, I can recommend you look at Edward Tufte's *The Visual Display of Quantitative Information* and Stephen Few's *Information Dashboard Design*. They both have other books on the same topic that are also very good.

As for the implementation, there are tools like Matplotlib and Seaborn in Python that can take you a long way. Normally data presentation is done in another tool though. Companies like SAP and IBM offer reporting solutions, while Tableau, Qlik, and Microsoft are a few of the bigger players in the graphical analytics space.

If you however just want to present the latest data in a mostly fixed fashion, Databricks actually has an offering. It's probably the strangest feature in their toolkit and rather bare-bones, but it can be useful in some cases.

The biggest advantage is that it's just there. When you've run your notebook, you can pick and choose what to put in a dashboard and just send a link to the recipient. When you rerun your code, everything will update. Let's take a look at this feature and how it looks for the end user.

Preparing data

Let's start by getting some data. We'll use the airlines data once again. It's not the greatest dataset for this as it actually requires cleaning, but we'll focus on the implementation rather than the actual result:

```
df = spark
      .read
      .format("csv")
      .option("inferSchema", "true")
      .option("header", "true")
      .load("/databricks-datasets/airlines/part-00000")
```

With the data in place, we run a simple query, just comparing three origin airports across the weekdays. We count the number of flights and set it as a column name. To get a sensible output, we also order it by DayOfWeek:

```
from pyspark.sql.functions import count

display(df
      .select('Origin','DayOfWeek')
      .filter(df.Origin.isin('LAS','SAN','OAK'))
      .groupBy('Origin','DayOfWeek')
      .agg(count('Origin').alias('NumberOfFlights'))
      .orderBy('DayOfWeek'))
```

Run the query and create a bar chart using DayOfWeek as keys, Origin as series grouping, and NumberOfFlights for values. This will give you three bars, one per airport, for every day in the week. That's all we'll do with this one for now:

```
from pyspark.sql.functions import count

display(df
      .select('UniqueCarrier','DayOfMonth')
      .filter(df.UniqueCarrier.isin('UA','PI'))
      .groupBy('UniqueCarrier','DayOfMonth')
      .count()
      .orderBy('DayOfMonth'))
```

This is another basic chart just showing the number of flights UA and PI did per day of the month across all the data. Make it into a line plot with DayOfMonth at the top, UniqueCarrier in the middle, and count in the bottom box. You should now have two charts ready in your notebook.

Using Matplotlib

If you're using Python and pandas, you'll sooner or later have to do some charting. When that happens, you'll most likely start using Matplotlib. It's one of the most powerful plotting tools available. Even though it's dense and hard to use, nothing has been able to dethrone it. So it's a good thing to learn and use. Unfortunately it doesn't currently play nicely with DataFrames.

If you want to use it with Pyspark, you need to do a bit of trickery. Basically you transform your DataFrame into a pandas dataframe. With that you can use all the normal syntax and get the results you expect. Here's an example:

```
df2pd = df.withColumn('arrdelayint', df.ArrDelay.cast("int"))
df2pd = df2pd
        .select('DayOfWeek','arrdelayint')
        .filter(df2.arrdelayint.between(-20,20)).na.drop()

import pandas as pd
import matplotlib.pyplot as plt

pddf = df2pd.toPandas()
fig, ax = plt.subplots()

pddf.boxplot(column=['arrdelayint'], by='DayOfWeek', ax=ax)

plt.suptitle('Boxplots')
ax.set_title('axes title')
ax.set_xlabel('Day of week')
ax.set_ylabel('Delay')

display()
```

We create a DataFrame with a new column, containing the arrival delays as an integer. We then filter out the small differences and drop the null rows. Next up are the import statements for pandas and matplotlib.

For this to work, we need a pandas dataframe, so we do a conversion. Then we do some basic matplotlib magic to create a boxplot of delays for every day of the week. The result will show, and you'll have no options in the cell as everything is done by code in this case.

Building and showing the dashboard

So now we have three nice graphs that we want the world to have access to. Time to create a dashboard. Click the View menu and select New Dashboard. This will take you to a new view where you'll see your data in a new way.

You can move things around, resize them, and remove the ones you don't want. If you do so, they'll only disappear from the dashboard, not the underlying notebook. You can also add them later if you change your mind.

When you're done playing around with the design, change the name to something other than Untitled and click Present Dashboard. This will show you the information in a clean design with an Update button in the upper-right corner.

The page you went to is also available as a link just below the dashboard name. You can copy that and send it to someone you want to look at the data. Let's see if we can create some interactivity before we do that though. Go back to the notebook by selecting code from the View menu.

Adding a widget

We've already come across the concept of widgets in this book when we passed arguments between notebooks. This time they will make more obvious sense. We'll use them to select data in our cells. Let's look at how that works:

```
carriers = df.select('UniqueCarrier').distinct().collect()

dbutils.widgets.dropdown("UniqueCarrierW"
        ,"UA"
        ,[str(c.UniqueCarrier) for c in carriers])
```

So we pull out all the carriers from the data and store them in a list. We then use that list when we create a dropdown widget called UniqueCarrierW. It'll contain all carriers, but the specified UA will be pre-selected. You'll see the resulting dropdown object just below the cluster list.

If you change the value in the list, nothing happens. The reason is there is no cell in the notebook that cares about the new widget. Let's create a graph that looks at what you selected. To do that we use the getArgument function:

```
df2 = df.withColumn('depdelayint', df.DepDelay.cast("int"))
df2 = df2.select('UniqueCarrier','depdelayint').na.drop()

display(df2
   .select('UniqueCarrier','depdelayint')
   .filter(df2.UniqueCarrier == getArgument('UniqueCarrierW'))
   .groupBy('UniqueCarrier')
   .avg())
```

In this case it actually matters what you selected. When you change your value, the cell will react and the data changes accordingly. This is pretty good if you want interactivity built in for users who are interested in what you do but don't really want to change anything you built. It works even better in a dashboard setting.

Adding a graph

If you create something in your notebook that you want to add to your dashboard, you can do that by clicking a button. Just to the right of the play button, in the upper-right corner of the active cell, you'll see a small chart button. Click it and a list of your dashboards will appear. Select where you want the result of the cell to show up.

Add the new cell with the widget-controlled result and create a new bar chart as well. Let's use basically the same data, but with a graphical touch. Run the following code and set the UniqueCarrier as keys and avg(depdelayint) for values:

```
df2 = df.withColumn('depdelayint', df.DepDelay.cast("int"))
df2 = df2.select('UniqueCarrier','depdelayint').na.drop()

display(df2.groupBy('UniqueCarrier').avg('depdelayint'))
```

Make sure to add this as well to your dashboard and then go back to the dashboard view. You'll notice that both new cells are in there and that you also have the widget at the top. Change it and the contents of the corresponding cell will change. You don't need to just use graphs, markdown text works as well.

Schedule run

The last thing we need to look at before leaving this topic is scheduling. When you are at the dashboard view, you have the option of asking Databricks to update all the cells that you have in your dashboard.

There is of course no difference in this compared to just scheduling the underlying notebook, but it might be nicer to connect the job to the view that is dependent on it. So let's do it.

Click the Schedule button in the upper-right corner. This will open a new toolbar where you'll see existing schedules. As we don't have any, you'll only see the blue New button. Click it and you'll get the standard cron window as we've seen earlier.

Select the time you want the data to be updated and click Ok. The schedule will appear. Click it to see all the job details. Especially be aware of which cluster setup will be spun up for this update.

REST API and Databricks

Most of the time you'll use Databricks through the graphical user interface. If you need a new cluster, you just go to the site and create one. The same goes for most other things you need to do. Once you go to a real environment where you need to do a lot of this however, manual steps aren't scalable. It's both tedious and error prone.

Fortunately there is a way to communicate with Databricks using code. They offer a RESTful application programming interface, API, with which you can control a lot of your environment. Note that you need a bought version of Databricks for this to work. The free community edition does not offer this functionality.

What you can do

What this means is that you can do stuff like spin up clusters, move files in and out of DBFS, and execute jobs, just to mention a few things. There are a number of APIs you can use, all containing a number of endpoints (which you can consider to be functions). If you want to automate any part of the Databricks work process, this is a gold mine.

You can find a list of all the APIs at the end of this section. Already now I can mention that there are APIs for the clusters, DBFS, the groups, the instance pooling, jobs, libraries, MLflow, SCIM, secrets, tokens, and workspaces. As you might realize, that's most of the stuff you use in Databricks.

CHAPTER 11   BITS AND PIECES

What you can't do

You can't control everything in Databricks through the APIs though. There are quite a few things that just don't work the way you'd like them to. For instance, you can't work with users. Actual users have to be handled either through the user interface or through an external solution with SCIM provisioning activated.

You also can't interact with the notebooks directly. You can get the files through the DBFS or export them, but not add a cell or get a result directly through the API. There are ways to get around this, but it's somewhat convoluted. So not everything is possible to do from afar using this interface.

Finally there is no way of doing work outside of Databricks. You might for instance want to automatically create workspaces for new projects. This cannot be done using the API. Instead you'll need to investigate what's available on Azure or AWS. Both have their own APIs for automating infrastructure work.

Still, the limitations are few, and there's often another way of doing what you want. Also, new features are added to the API every now and then so just because it isn't there today doesn't mean it won't be there tomorrow.

Getting ready for APIs

What you need to get started is a token and the domain you are using. The domain is easy to figure out. Just look at the URL you're using when in Databricks. It'll be something like `https://westeurope.azuredatabricks.net`. It'll be slightly different if you're in another data center or if you are using AWS.

Next up, you need the token. We've talked about this earlier, but let's walk through it again. Click the little icon in the upper-right corner and go to the User Settings view. Make sure you're on the Access Token tab and click the Generate New Token button. Type in a description and generate the token. Copy it and store it somewhere you can access it. We'll be using it multiple times in the next few pages.

Next up, you need an environment to run the code in. We'll be using Python, but as it's a normal RESTful API, you can use pretty much whatever you want, even a tool like Postman. I do recommend using something in which you can write real code. If you want, you can even use a notebook in Databricks.

Before we get into the examples, let me give you a quick run-though on how you work with the API. With the basics defined, you just have to pick the endpoint you're interested in, set the correct parameters, and then either get or post data, depending on whether you pass on data or not. It's very easy once you get the hang of it.

Example: Get cluster data

Let's look at a simple example. That's usually the easiest way to understand what's happening. If you've worked with web services before, you'll quickly recognize the core pattern. Set a header, define an endpoint, and just send a request.

Note by the way that I'm hard-coding tokens here. You shouldn't do that in a real environment. Use some type of obfuscation tool, like Databricks Secrets, to make sure no one else can access the API using your credentials.:

```
import requests, base64

token = b"<yourtoken>"
domain = "https://westeurope.azuredatabricks.net/api/2.0"
endPoint = domain + "/clusters/list"

header = {"Authorization": b"Basic " + base64.standard_b64encode(b"token:" + token)}

res = requests.get(endPoint, headers=header)

print(res.text)
```

We start up importing two libraries. The first one is a HTTP library we use to send requests over the Web. The other one, base64, is needed to encode the token string. By using the b-notation, we create a string of 8-bit bytes, securing that all characters are transmitted correctly.

Next up, we define the token, domain, and endpoint. We then create a header. With all parts in place, we do a call to the API, returning the result to the res variable. Finally we just dump the result, which is all the clusters with a lot of extra data.

You might notice that the result is returned in a JSON format. If we handle it correctly, we can actually parse the contents and use it to do logic or just present data in a nicer way. With this you can do some interesting things.

CHAPTER 11 BITS AND PIECES

One thing might for instance be to see how much money is being lost by interactive clusters not being shut down properly. Even if you have an auto-shutdown feature, it can roll up into real money if you use the default of 2 hours.

If you, for instance, work for 6 hours every day and don't shut down the clusters, up to 25 percent will be spent on idling (assuming to scaling). Let's look at what that code could look like, not taking scaling into account. Note that I just add prices for the clusters I use. Feel free to add the ones you are using:

```
import requests, json, base64, datetime

token = b"<yourtoken>"
domain = "https://westeurope.azuredatabricks.net/api/2.0"

prices = { "DS3" : 0.572
         ,"DS4" : 1.144
         ,"DS5" : 2.287
         ,"D32s" : 4.32 }

header = {"Authorization": b"Basic " + base64.standard_b64encode(b"token:" + token)}

endPoint = domain + "/clusters/list"

res = requests.get(endPoint, headers=header)

js = json.loads(res.text)

print("Poorly handled clusters:")
for k in js['clusters']:
  source = k['cluster_source']
  state = k['state']
   if (source != 'JOB')&(state != 'RUNNING'):
   termtime = k['terminated_time']
   tdate = datetime.datetime.fromtimestamp(int(termtime)/1000).date()
   if tdate == datetime.datetime.today().date():
    typ = k['node_type_id']

    name = k['cluster_name']
    mins = k['autotermination_minutes']
    minwork = k['autoscale']['min_workers']
```

```
    source = k['cluster_source']
    totnodes = int(minwork) + 1
    totmins = totnodes * int(mins)
    typematch =  typ.split('_')[1]
    pricepermin = prices[typematch] / 60
    totcost = pricepermin * totmins
    if state == 'TERMINATED':
     reason = k['termination_reason']['code']
     if reason == 'INACTIVITY':
      print("{} ({}) with {} nodes ran for {} minutes unnecessarily.
      Price: ${}").format(name, typ, totnodes, totmins, totcost)
```

This is a slightly cumbersome way to do this, but it will be easy to read I hope. The start of the script is the same as in the preceding code. We just get a list of all the clusters. Here we take the result and turn it into a JSON object using the loads function. The only thing we added is a dictionary of prices for different cluster node types.

Note that it's a good idea to wrap code like this inside try/except clauses. That makes sure you don't get an error due to an unexpected state of the cluster. For instance, you won't get all the values back for a running cluster.

With all that done, it's time to look through all the clusters, looking for interactive ones by excluding the job clusters. We also look at the termination time to only pick up clusters that finished today.

If these requirements are met, we pull out some data for the cluster. We check stuff like the auto-termination time, the minimum number of nodes running, and the minute price. We then have all the base data we need.

Finally we check to see if the cluster is currently terminated and if it was killed due to inactivity. If so, we print out the minimum cost of the idle time. Note that the actual cost could very well be higher as the idle period could start with more nodes running. Still, this is a good reminder to email developers who forget to shut down huge clusters.

Example: Set up and execute a job

Working with clusters is a common operation. Jobs might be even more common, especially if you are running in production from an external scheduler. The way you'd do that is through the API. First, create a notebook and call it TestRun. Just add a simple sleep command:

CHAPTER 11 BITS AND PIECES

```
from time import sleep
sleep(100)
```

Next up, let's create a job template to see what it looks like. The first thing we need to do is to define what it is we want to run. We do this by creating a JSON structure with all the details needed. We don't actually need the library, but I'll add it so you see how to do it:

```
import requests, json, base64

job_json = {
 "name": "MyJob",
 "new_cluster": {
  "spark_version": "5.5.x-scala2.11",
  "node_type_id": "Standard_DS3_v2",
  "num_workers": 4
 },
 "libraries": [ {
  "jar": "dbfs:/FileStore/jars/abcd123_aa91_4f96_abcd_a8d19cc39a28-ojdbc8.jar"
 }],
 "notebook_task": {
  "notebook_path": "/Users/robert.ilijason/TestRun"
 }
}
```

As you can see, we tell Databricks we want to use a temporary job cluster using a given Spark version and node type with four worker nodes. We also load a jar file. Finally we define which notebook to run. This is just the definition though. We now need to post it into Databricks using the API:

```
token = b"<yourtoken>"
domain = "https://westeurope.azuredatabricks.net/api/2.0"

header = {"Authorization": b"Basic " + base64.standard_b64encode(b"token:" + token)}
endPoint = domain + "/jobs/create"
```

CHAPTER 11 BITS AND PIECES

```
res = requests.post(endPoint
        ,headers=header
        ,data=json.dumps(job_json))

print(res.text)
```

You'll notice that it looks much like the code we used for the clusters. This is true for most of the API. The biggest difference here is that we also send data. It needs to be JSONified, so we use the dumps function. If everything goes well, you'll get an identifier, a job id, back.

With this we have defined a job and loaded it into Databricks. If you look at the Jobs view, you'll find it there. We still haven't executed it though. That can also be done with the API, so let's do just that:

```
import requests, json, base64

job_json = {
 "job_id": <the Job ID from the last job>
}
token = b"<yourtoken>"
domain = "https://westeurope.azuredatabricks.net/api/2.0"

header = {"Authorization": b"Basic " + base64.standard_b64encode(b"token:" + token)}
endPoint = endPoint = domain+"/jobs/run-now"

res = requests.post(endPoint
        ,headers=header
        ,data=json.dumps(job_json))

print(res.text)
```

Now the job is up and running, thanks to the run-now endpoint. The command will start the job, but not wait for it to finish. Instead you get a run identifier. With this one you can track the status if you want:

```
import requests, json, base64

job_json = {
 "run_id": <the Run ID from the last job>
}
```

```
token = b"<yourtoken>"
domain = "https://westeurope.azuredatabricks.net/api/2.0"

header = {"Authorization": b"Basic " + base64.standard_b64encode(b"token:" 
+ token)}
endPoint = endPoint = domain+"/jobs/runs/get"

res = requests.get(endPoint
        ,headers=header
        ,data=json.dumps(job_json))
js = json.loads(res.text)

print(js['state']['state_message'])
```

If you run this just after you started the job, you'll probably get a Waiting for cluster message, but after a while it'll start running and finally finish ok. If you want to see the different states, you can increase the sleep period for as long as you need.

Example: Get the notebooks

If you want to use version control, there's a built-in feature for that in Databricks. For a number of reasons, you might want to do this in a different way though. To do that you need to get the notebooks out of the system – possibly also back in.

Let's look at how to export a notebook. You can use any of the SOURCE, HTML, JUPYTER, and DBC formats. In our case we want to look at the actual data, so we'll go with source. If you want to do a straight clone, other options might be better:

```
import requests, json, base64

job_json = {
  "path": "/Users/robert.ilijasson/TestRun",
  "format": "SOURCE"
}

token = b"<yourtoken>"
domain = "https://westeurope.azuredatabricks.net/api/2.0"

header = {"Authorization": b"Basic " + base64.standard_b64encode(b"token:" 
+ token)}
endPoint = domain + "/workspace/export"
```

```
res = requests.get(endPoint
      ,headers=header
      ,data=json.dumps(job_json))
js = json.loads(res.text)

print(base64.b64decode(js['content']).decode('ascii'))
```

This should be clear by now. We set our parameters, send them to Databricks, and get a result. We parse the contents, and as we get it encoded, we need to run it through the base64 decoding. The result is the code in the notebook, ready for checking in if you so choose.

All the APIs and what they do

We've so far just looked at a few of the many options you have through the API. They all have more or less the same structure though. Just look at what parameters are needed, make sure you use the correct HTTP method, and go.

The following is a list of all the APIs in version 2.0. If you want more detailed information about them and all the endpoints, check out the documentation. You'll find it at Databricks's home page – `https://docs.databricks.com/dev-tools/api/latest/index.html`:

Clusters API gives you the ability to manipulate the clusters. You can, for instance, run commands like create, delete, start, stop, resize, and much more. This is a crucial API if you want to automate creation and keep track of what's happening in your system.

DBFS API lets you work with the Databricks File System. If you want to create a directory, list contents, or move files, it's all possible. You can also use it to upload and download files to and from the file system.

Groups API is what you'll use to set up groups from the command prompt. You can also list who's in the group, add new users, and remove the existing ones. Most environments will probably use a different solution for this, but it's there if you need it.

Instance Pools API is relevant if you use the cluster pooling feature. You can create new ones, delete them, and list what exists. This will probably be a part of your automated environment setup.

Instance Profiles API is available for AWS users. You can add and remove the instance profiles that users can start clusters with. There's also, of course, a list command to see what's already set up.

CHAPTER 11 BITS AND PIECES

Jobs API is probably the one you'll use the most. We've looked at it a little bit in the examples earlier, but there's a lot more in there. Most of it relates to the job executions, or runs. You might, for instance, want to get the run results, which is possible.

Libraries API is a simple way to add or remove libraries. If you want to update a library on a large number of workspaces, this API will make it easier. Unfortunately the cluster you install on has to be up and running, or the command will fail. Use the Clusters API to spin it up first.

MLflow API is a large API that is a managed version of the MLflow tracking server. There's a lot going on in the package, and you can control most of it. For instance, you can interact with experiments, runs, and models.

SCIM API can be used for managing users and groups through the System for Cross-domain Identity Management (SCIM) standard. This is in a public preview at the time of this writing, but it promises to give you administrative powers if you use the SCIM provisioning feature.

Secrets API we looked at already. This will give you the ability to hide passwords, tokens, and other sensitive information in notebooks. While this is a great feature, consider using something that works across multiple systems.

Token API is useful if you need a temporary token. You can create, revoke, and list the existing ones. You do, of course, need to have a token already to call the endpoint, so it's mostly for short-term uses.

Workspace API lets you export and import notebooks. You can also use it to create folders and list contents in the workspace. There's also a useful get-status endpoint that will tell you if a resource exists. It simplifies scripting a bit.

I should mention that at the time of this writing, version 1.2 of the API still exists and actually contains a little bit of extra functionality. Most importantly, you can execute individual commands and JAR files. So in theory you can run Databricks just as a Spark front end. I don't think it's a great experience, but know it's there if you need it.

Delta streaming

This book has been about running large-scale analytics in a traditional way. Load data in bulk, clean it up, and run some algorithm at the end. With data coming in faster and decisions having to be done quickly, streaming has become popular, especially in the ecommerce field.

The idea with streaming is that you consume the data continuously and run your aggregations, checks, and algorithms on the fly with minimal lag. If you have a lot of customers and want to adjust your prices on a minute-to-minute basis, you can't rely on batch-oriented solutions.

Streaming does require a bit more of the data flow as you don't have much time to handle errors. It's just not as resilient as traditional, batch-oriented solutions where you can secure the data before you move to the next step. Validate even more is a general tip. You don't want to make decisions on bad data.

While streaming is hot right now, it doesn't make sense in every use case. Quite the opposite actually. It's unnecessary to update reports every second if the result is consumed daily or weekly. Most important decisions are not made in a split second, even though the hype about data-driven organizations might make you think so.

That said, you might want to use streaming to minimize load times, to make integrations easier, or for a number of other reasons. So understanding how it works is helpful, which is why we'll end with this – an open door toward where you could go next.

Note though that streaming is a big topic and could easily cover a book as thick as this one, so we're just scratching the surface here. If you find it interesting, I'd recommend reading up on Apache Kafka and then continue the work on Databricks using Delta Lake.

Running a stream

Let's set up a streaming test. We'll create some dummy data, activate streaming, and consume it. This will give you a sense of the ideas around streaming. To make it easier for ourselves, we'll use plain text files as the source. In reality it is probably more likely you'll get the information from an external tool, like Apache Kafka:

```
dbutils.fs.mkdirs('/streamtst')
```

We start with creating a folder where we'll store our data files. Note that we use dbutils and that it ends up on DBFS. Next up, we need to create something to process. Let's dump a simple structure in JSON format into a text file:

```
import json
import random
import time
```

CHAPTER 11 BITS AND PIECES

```
d = {'unix_timestamp':time.time()
    ,'random_number':random.randint(1,10)}

with open('/dbfs/streamtst/0.json', 'w') as f:
    json.dump(d, f)

%fs ls /streamtst/
```

We create a simple dictionary with two columns. The first, unix_timestamp, is populated with the current epoch time. A random number fills the second column, random_number. We convert the result to JSON format and dump the result into our folder. To make all this possible, we import three libraries.

The file system command will list everything in the streamtst folder. There should only be one file, our 0.json, there. Next up, we need to define the schema so Databricks knows what we are working with:

```
from pyspark.sql.types import TimestampType, IntegerType, StructType, StructField

schema = StructType([
        StructField("unix_timestamp", TimestampType(), True),
StructField("random_number", IntegerType(), True) ])
```

We've done this before, so nothing new here. We import the types we're using and create the schema manually. With this we can create the readStream, which works like a connection. So as with transformations, nothing will actually happen when you run the command:

```
dfin = (spark
        .readStream
        .schema(schema)
        .option('maxFilesPerTrigger',1)
        .json('/streamtst/'))
```

Nothing exciting here. We define it's a readStream and provide the schema and the source folder. The maxFilesPerTrigger tells Databricks to just process one file at every refresh. If you remove it, it'll use the default of 1,000. We'll look at what that means further into this chapter.

Now we'll define what to do with the incoming data. You can choose to just pass the data along, but let's do something slightly more interesting: a simple aggregation where we count the number of occurrences of each random number, for instance:

```
dfstream = (dfin.groupBy(dfin.random_number).count())
```

Now we have everything defined. Time to start the writeStream and get everything up and running for real. We'll create a table called random_numbers in memory to begin with. The outputMode parameter defines what will be sent along into the so-called sink. You can choose between append, update, and complete.

Append will only add only new rows, Update will rewrite modified rows, and Complete handles everything every time. The latter is useful in our case as we want to run an aggregation on the data:

```
dfrun = (
  dfstream
    .writeStream
    .format("memory")
    .queryName("random_numbers")
    .outputMode("complete")
    .start()
)
```

Running this will result in a slightly different output than you are used to. First, you'll see Databricks working with a text that says Streams initializing. It means exactly what it says. After a few seconds, it'll start properly and it'll work with the data in our streamtst folder. You know it's ready once you see a green icon and the text random_numbers to the right.

Another thing that's different is that you can actually continue to write code, even though that cell that's streaming is running. So once it's running, execute the following query to see that it worked:

```
%sql select * from random_numbers;
```

If everything went well, you'll get the data from the file we created. Great that it worked, but not too exciting. Let's see what happens if we add another file to our folder. To do that we just run the same code as before, but with a new filename:

```
d = {'unix_timestamp':time.time()
    ,'random_number':random.randint(1,10)}

with open('/dbfs/streamtst/1.json', 'w') as f:
    json.dump(d, f)
```

Then wait for the stream to pick it up and then run the query again. You'll notice a message telling you how fresh the data is. Once it's done try to run the same query as before. You'll notice that you now have two rows:

```
%sql select * from random_numbers;
```

As you'll see the data is picked up automatically. That's the magic. You can just add data and it'll automatically be handled. Let's make this a bit more interesting. Create a plot with random_number for Keys and count for Values. Then let's add more files:

```
for i in range(2,100):
  d = {'unix_timestamp':time.time()
      ,'random_number':random.randint(1,10)}

  with open('/dbfs/streamtst/{}.json'.format(i), 'w') as f:
      json.dump(d, f)
```

```
%fs ls /streamtst/
```

You'll see that we now have a bunch of files in the folder. Let's see how Databricks handles our data. You can just rerun the query with the plot. Execute it every minute or so, and you'll see how the information changes.

The files are added to the in-memory table one file at a time. That's because we've defined that using the maxFilesPerTrigger. If we don't use that, we'll get a slightly different experience. First, let's see how we stop the stream.

Checking and stopping the streams

While it's good to have the streaming process running if you get data, it does keep your cluster up and running. So it can turn expensive if you're not careful. If you want to see if any streams are running, you can do so by running this code.

```
for stream in spark.streams.active:
  print("{}, {}".format(stream.name, stream.id))
```

You'll notice that we have one running. Now, if you want to stop it. you'll have to run another command. As we use the dfrun name for the active stream, that's what we'll stop and we do it like this:

```
dfrun.stop()
```

If you want to stop all active streams, you can just run a modified version of the listing we did earlier. The following command will do exactly that. Just be careful not to stop anything by mistake:

```
for stream in spark.streams.active:
  stream.stop()
```

An interesting detail is that the table is still alive. It won't be updated anymore, but you can query it as long as you don't stop the cluster or actively remove it. Let's verify it by running the same cell with the plot once again:

```
%sql select * from random_numbers;
```

Running it faster

As mentioned we limited the speed in our example. That's not what you'd normally do. Instead you'd probably let Databricks run at a slightly higher pace. What you get without any setting at all is 1000. Let's try it:

```
dfin = (spark.readStream.schema(schema).json('/streamtst/'))
dfstream = (dfin.groupBy(dfin.random_number).count())
dfrun = (
  dfstream
    .writeStream
    .format("memory")
    .queryName("random_numbers")
    .outputMode("complete")
    .start()
)
```

Once it's initialized, which will take a bit longer this time, run the query with the plot once again. You'll notice that you have much higher bars immediately. It processed all of our data immediately. Suddenly you can throw a lot more at the stream as it'll handle it in bulk. Feel free to create many more files and try it out.

CHAPTER 11 BITS AND PIECES

Using checkpoints

So far we've done all our streaming in memory. That means you'll start from scratch every time you start your processing. That's ok in some cases, but often you want to continue where you left off. You might also want to materialize the information into a table. Let's look at how to do that:

```
dbutils.fs.mkdirs('/streamtstchk/ )
```

We'll start by creating a new folder for our checkpoint files. Checkpoints are a kind of a ledger Apache Spark uses to remember what happened. They will keep track of our status and make sure we can pick up where we left off. Next, let's set things up like before:

```
dfin = (spark
        .readStream
        .schema(schema)
        .option('maxFilesPerTrigger',1)
        .json('/streamtst/'))

dfstream = (dfin.groupBy(dfin.random_number).count())

dfrun = (
  dfstream
    .writeStream
    .format("delta")
    .option("checkpointLocation", '/streamtstchk/_checkpoint')
    .queryName("random_numbers")
    .outputMode("complete")
    .table('stest')
)
```

The last part is where things look a little bit different. We tell Databricks where we want to keep the checkpoint files; and we also define a table, stest, where data will go. Run the code, and once the stream runs, you can run the following query to keep track of the progress:

```
%sql select * from stest;
```

You'll see that we see the rows slowly building up, since we limited the speed using the `maxFilesPerTrigger` setting. Wait for a few rows to build up and then stop the active stream using the stop command. When it finishes, start it again and run the query to see the contents:

```
dfrun.stop()

dfrun = (
  dfstream
    .writeStream
    .format("delta")
    .option("checkpointLocation", '/streamtstchk/_checkpoint')
    .queryName("random_numbers")
    .outputMode("complete")
    .table('stest')
)
```

You'll notice that you already have data in the table. The processing will continue where it finished, and you don't have to reprocess information that you already handled. This way you can work with your data for a while, shut the system down when you're done, and continue the next day.

Index

A

Access control
    cluster/pool/jobs, 223
    personal, 225
    table, 223, 224
    workspace, 222
Access Control Lists (ACLs), 219
agg function, 119
ALTER command, 93
Amazon Web Service (AWS), 32
Apache Spark
    Apache Software Foundation, 17
    architecture (*see* Spark architecture)
    cluster/parallel processing solutions, 15
    components, 24
    Databricks moon, 18
    data challenges, 17
    definition, 15
    large-scale analytics, 17
    scalability, 16
    SQL and DataFrame, 16
Apache Spark DataFrame, 55, 103, 109, 244
Apache Spark/traditional software, 104
approxQuantile, 151
Apress, 218
Authentication Mechanism, 181

B

Bits and pieces
    creating data, 229, 230
    frequent pattern growth, 228, 229
    MLlib, 228
    parsing results, 232
    preparing data, 230, 231
    running algorithm, 231

C

Cleaning/transforming data
    caching data, 158–160
    columns, 153, 154
    createDataFrame procedure, 142
    data compression, 160, 162, 163
    explode command, 156
    extreme values, 150–152
    fillna command, 146, 147
    isNull, 143, 144
    isNull, dropna, 144, 145
    lambda functions, 163–165
    lazy Pyspark, 156–158
    pivoting, 154, 155
    removing duplicates, 148, 149
Cloud
    commercial edition
        Amazon Web Services and Microsoft Azure, 30

INDEX

Cloud (*cont.*)
    AWS, 32, 33, 35
    Azure Databricks, 36, 37
    Data Engineering, 31
    DBU, 32
    community edition
        checkbox, 29, 30
        GET STARTED button, 29
        6-gigabyte driver node, 28
        version, 28
    computing, 9
    enterprise companies, 27
Clusters, 42–44
collect_list function, 231
collect_set function, 231
combo commands
    agg function, 119, 120
    alias command, 119
    & character, 117
    build chains, 116
    built-in command, 122, 123
    classical loops, 124
    DataFrame, 117, 119, 123–125
    date_format function, 121, 122
    lit function, 125
    multiple filters, 117
    orderBy, 118
    pyspark modules, 116
Command line interface (CLI), 23
    ACLs, 219
    cluster commands, 212, 213
    creating/running jobs, 214, 215
    Databricks File System accessing, 215, 216
    picking up notebooks, 216, 217
    secrets, 217, 218
    setting up, 212, 213
cron command, 206
cx_oracle library, 196

D

Data Analytics, 31
Databricks, 1, 14
    access (*see* Access control)
    accessing external systems, 193
    cloud solutions (*see* Cloud)
    clusters, 178
    code operating system, 202, 203
    connecting external systems
        Azure SQL, 195
        MongoDB, 198, 199
        Oracle, 196–198
    connecting tools
        Microsoft Excel, 185
        Microsoft's Power BI, 186, 187
        PyCharm, 188–190
        Tableau OS X, 187
    connection, 178
    handle errors, 202
    job
        Cluster information line, 204
        concurrency option, 205
        notebooks, 205–208
        parameters, 210, 211
        scheduling, 206
        widgets, 208, 209
    libraries, 194
    re-runnable code, 202
    revisiting cost, 220
    RStudio Server, 191–193
    SCIM provisioning, 222
    Tableau, 177
    users/groups, 221, 222
Databricks File System (DFS), 215
    Apache Spark cluster, 51, 52
    constructor_results, 61, 62
    data available, 45, 46
    datasets, 53

INDEX

dbutils package, 53, 54
FileStore folder, 55, 71, 72
Github repository, 60
Hive Metastore, 56
Linux operating system, 52
mount points (*see* Mount file system)
notebook user interface, 71
README.md file, 53
schemas/databases/tables, 55
source files, types
    binary, 59
    CSV, 57
    delimited, 57
    JSON, 58
    transportation, 60
UI, 39–42
web, getting data (*see* Web)
Databricks Unit (DBU), 32
Data Manipulation Language (DML), 94
Data source name (DSN), 183
date_format function, 121
DB_IS_DRIVER, 198
dbutils command, 131
dbutils.fs.unmount command, 70
dbutils package, 54
Delta Lake SQL
    metadata selection, 99, 100
    optimization command, 98, 99
    rediscovered databases, 95, 96
    statistics, 101
    transaction logs, 99
    UPDATE/INSERT/DELETE, 97
Delta streaming
    batch-oriented solutions, 261
    checking/sropping streams, 264, 265
    dbutils, 261
    large-scale analytics, 260
    outputMode parameter, 263
    random_numbers, 263, 264
    readStream, 262
    unix_timestamp, 262
    using checkpoints, 266, 267
DESCRIBE command, 92
Directed Acyclic Graph (DAG), 141
display command, 110
Driver Type, 44
dropTempView command, 128

E

exceptAll command, 133
explode command, 156
Extraction, 139
Extract/Transform/Load (ETL), 140, 142
    cleaning and transforming (*see* Cleaning/transforming data)
    storing and shuffling (*see* Storing/shuffling data)

F

fillna command, 146
FP-Growth algorithm, 230
FPGrowth library, 231
Fueling analytics, 5, 6

G

getArgument function, 250
getNumPartitions command, 170
groupBy function, 116

H

Hadoop Distributed File System, 51

I

inputCols, 147
INSERT command, 94
INTERSECT command, 83
isNull, 143

J

Java Database Connectivity (JDBC), 72, 178
JavaScript Object Notation (JSON), 16, 58, 90

K

Koalas, 227, 242
Koalas, Pandas
    DataFrame, 243–245
    future, 246
    preparing data, 247
    presenting data, 246
    Spark, 242

L

Large scale data analytics
    analyse data, 10, 11
    Big data, 4, 5
    Cambridge analytica, target, 13
    clouds, 8, 9
    customer analytics, target, 13
    databricks, 9, 10
    datasets, 3
    data warehouse, 3
    intense analytics processing, 3
    key drivers, 2
    Telematics, Volvo Trucks, 12
    Visa, 12

Lempel-Ziv-Oberhumer (LZO), 163
LIKE command, 101

M

Machine Learning library (MLib), 228
Main.json, 214
mapDateToTimestamp, 238
Massively Parallel Processing (MPP), 19
Matplotlib
    adding graph, 250, 251
    dashboard, 249
    Pyspark, 248
    widgets, 249, 250
Microsoft Azure/Amazon Web Services (AWS), 9
Microsoft's Power BI, 186
MLflow, 227
    checking results, 236
    databricks, connect, 238
    original tables, 237
    pulling data, 239, 240
    running code, 233–235
    tracking/project packaging, 233
    update table, 236, 237, 241
    verifying formats, 240, 241
MLlib, 227
MongoDB, 198
Mount file system
    Amazon S3, 67, 68
    Microsoft blob storage, 69, 70
    unmount command, 70

N

namelist() function, 65
Notebooks, 46–49

INDEX

O

ODBC and JDBC
    create test table, 180, 181
    creating token, 179
    drivers, 178
    OS X, 182–184
    preparing cluster, 180
    Simba's version, 178
    Windows setting up, 181
Open Database Connectivity (ODBC), 72, 178
Open Source Software (OSS), 2
Optimized Row Columnar (ORC), 16, 58

P, Q

Pandas, 242
Pay-as-you-go model, 17
printSchema command, 110
pyspark.sql.types library, 112
Python
    branching, 105
    DataFrames
        bins function, 127
        CSV, 132
        Databricks, 130
        dbutils package, 130
        format command, 132
        saveAsTable, 129
        StringType, 127
        subset, 125
        temporary view, 128
        user-defined function, 126
    DataFrames API, 108, 109, 114, 115
    dataset, 132–136
    date format patterns, 120, 121
    finding data, 107, 108
    functions, 106
    getting data, 110–113
    languages, 103
    lines, 106
    loops, 105
    Nulls, 115
    strings, 105, 107

R

Resilient Distributed Dataset (RDD), 21
REST API
    Access Token tab, 252
    Azure/AWS, 252
    cluster data, 253–255, 259
    DBFS, 251, 252, 259
    groups, 259
    instance pools, 259
    instance profiles, 259
    jobs, 260
    set up/execute job, 255–258
Row-based and column-based storage, 59
RStudio Server, 191

S

%sh magical command, 62
Simba Spark drivers, 179
Spark architecture
    Clusters, 19
    data processing, 22
    MPP, 19
    processing, 20
    RDD, 21, 25
    servers, 19
    storing data, 23
Spark driver, 72

INDEX

Spark GraphX, 24
Spark MLlib, 24
Spark SQL
    creating data, 91–93
    database, 76, 77
    databricks flavor, 75, 76
    DML, 94, 95
    filtering data, 79–82
    functions, 85, 86
    hierarchical data, 90, 91
    joins/merges, 82–84
    ordering data, 84
    picking up data, 77, 78
    stored query, 89
    Windowing functions, 86–89
Spark UI
    DAG, 141
    Environment tab, 141
    event timeline, 140
    globbing paths, 141
splits argument, 152
stack command, 155
Storing/shuffling data
    managed *vs.* unmanaged
        tables, 167, 168
    partitions, 169–174
    save modes, 165–167
Structured Query Language
    (SQL), 102
surrogateDF, 147
symmetric_difference
    function, 241

T

Tableau, 177
Transformation, 140

U

UDFs and lambda functions, 163–165
UNION ALL command, 82
union command, 133
UNION statement, 82
UniqueCarrierW, 249
unmount command, 70
UPDATE command, 97
user-defined function, 126

V

VectorAssembler, 234

W, X, Y

Web
    Python, 64–66
    shell, 62–64
    SQL, 66, 67
wget command, 62
withColumn command, 115
withColumnRenamed function, 135
Worker Type, 44

Z

ZipFile function, 65

Printed in Great Britain
by Amazon